The Symphony of Space and Time: A Multi-Scaled Approach to Wildlife Habitat

Kinky

Copyright © 2024 by Kinky

All rights reserved. No part of this book may be reproduced in any man-ner whatsoever without written permission except in the case of brief quotations embodied in critical articles and reviews.
First Printing, 2024

Table of Contents

Chapter 1: Variation in behavior drives multiscale responses to habitat conditions in timber

rattlesnakes (*Crotalus horridus*)

Introduction

Organisms respond to environmental variation simultaneously across numerous temporal and spatial scales. Understanding habitat use across these scales is key to developing appropriate management strategies for rare or threatened wildlife. Modern telemetry and GIS technology allow for the collection of increasingly high-resolution data and can pair well with traditional point selection function (PSF) analyses (Boyce et al. 2002, Zeller et al. 2014). More recently, an increasing number of studies have stressed the importance of scale in such analyses and have highlighted that few studies empirically select appropriate scales for habitat models (Holland et al. 2004, Smith et al. 2011, Jackson and Fahrig 2015, McGarigal et al. 2016). Empirical testing of appropriate scales, especially the grain of environmental variables, is critical given the recent proliferation of and ready access to remotely sensed data that can help explain wildlife site use (Thompson and McGarigal 2002, Neumann et al. 2015).

The appropriate spatial extent at which to measure habitat features is the "scale of effect" (Jackson and Fahrig 2012) and depends on species' traits and the biological variable of interest (Miguet et al., 2016). Miguet et al. (2016) discuss a framework for assessing the scale of effect for different variables that might influence a species' space use. The authors assert that biological processes occurring over longer time periods (e.g., occurrence) will have larger spatial scales of effect than those happening over short time periods (e.g., fecundity). Although relatively few studies have found support for this hypothesis (Cushman and McGarigal 2004, Jackson and Fahrig 2014), the general idea of a link between temporal and spatial scales is a foundational principal of landscape ecology (Peterson et al. 1998).

A temporal-spatial scale connection should be evident within the temporally variable behavioral patterns of wildlife. For ectotherms, site-specific behavior is often related to thermoregulation, which drives much of their habitat selection (Stevenson 1985, Huey 1991).

Both the thermal landscape and energetic requirements of ectotherms fluctuate seasonally,

contributing to variation in habitat use throughout the year (Waldron et al. 2006, George et al.

2017). However, individuals within a single population often operate under different selective

pressures at the same time of year, owing to variation in body condition or reproductive state

(Lesmerises and St-Laurent 2017). Thus, generalizing population-level, seasonal shifts in

behavior and site use may overlook important, fine-scale heterogeneity.

Behavioral variation is often difficult to assess when direct observation is infrequent or

may disrupt natural behaviors. Rattlesnakes are ideal subjects with which to study the effect

behavior has on habitat selection as they can be directly observed using radio telemetry with

minimal disturbance (Brown et al. 1982, Reinert and Zappalorti 1988) and they exhibit

stereotyped behaviors (Clark 2004, Reinert et al. 2011a). While previous studies have detailed

the effect that gestation has on space use in snakes (Charland and Gregory 1995, Reinert and

Zappalorti 1988, Sprague and Bateman 2018), other behaviors have been largely ignored.

Waldron et al. (2006) identified "seasons" based on foraging, breeding, and hibernation to

describe space use in timber rattlesnakes (*Crotalus horridus*) but did not address changes in

behavior within a season.

Rattlesnakes are ovoviviparous with long (4–5 months), energetically-costly gestation

periods (Ernst and Ernst 2003). Females reduce movements, seek out relatively warmer or more

exposed sites than would normally be used, and decrease or stop foraging for food (Brown 1991,

Martin 1993, Gardner-Santana and Beaupre 2009). Ecdysis and digestion occur over a shorter

timeframe (7–14 days and 3–7 days, respectively) but these physiological states can also drive

snakes to select more open habitats for thermoregulation (Greenwald and Kanter 1979, Semlitsch

1979). Foraging behavior may be even more sporadic and opportunistic, with associated changes in behavior during a single day (Clark 2004, Reinert et al. 2011a). Snakes making opportunistic site use decisions related to foraging and digestion may consequently select habitat at finer scales than snakes making decisions related to more predictable, long-term behavioral states (i.e., gestation and ecdysis). If this is true, we would expect gestating females to use different macrohabitats than non-gravid females (e.g., early successional parcels instead of mature forests). Conversely, foraging snakes would more opportunistically vary site use based on local or microhabitat features. We would expect snakes in ecdysis or those digesting a recent meal to represent an intermediate trend, either selecting features based on an intermediate spatial scale or exhibiting a mix of macrohabitat and microhabitat-level selection.

We tested whether predictable, long-term behavior patterns have a larger scale of effect when compared to more sporadic, short-term behaviors in timber rattlesnakes. We considered a multi-scale framework to determine the appropriate scale of effect of several geospatial landscape features on four behaviors of interest in wild timber rattlesnakes. We hypothesized that snakes making short-term decisions (i.e., foraging or digestion) on site use would select habitat at finer scales than snakes making decisions related to predictable, long-term behaviors (i.e. gestation or ecdysis).

Methods

Study Site

Vinton Furnace State Experimental Forest (VFSF) is a 4,892-ha property located in Vinton County, Ohio, USA, and within the Southern Unglaciated Allegheny Plateau Ecological section (Cleland et al. 2007). The landscape features a dissected topography, including sharp

4

ridges and valleys with relatively low relief (~100 m). The surrounding region is primarily

forested, consisting of mostly second-growth stands recovering from heavy exploitation during

the mid-to-late 1800s (Stout, 1933). The mixed-mesophytic forest type of the region is primarily

dominated by oak species (*Quercus* spp.), especially on dry ridgetops and southwestern

hillslopes, and transitions to mesophytic forest assemblages, including sugar maple (*Acer

saccharum*), American beech (*Fagus grandifolia*), and yellow poplar (*Liriodendron tulipifera*),

on opposing northeastern hillslopes and bottomlands (Hix and Pearcy 1997; Adams et al. 2019).

The study site also includes small tracts of pine plantations, usually composed of monospecific

stands of *Pinus strobus*, *P. resinosa*, or *P. echinata*, while other native conifers, including *P.

virginiana*, *P. rigida*, and *Tsuga canadensis*, combine at low densities with deciduous

hardwoods. In addition, the study site incorporates active, ongoing forest management,

especially within the central Raccoon Ecological Management Area. Varying management

approaches, including even- and uneven-aged silvicultural prescriptions, as well as prescribed

fire, promote heterogeneity in habitat conditions within the study site (Ducey 1982).

Telemetry

We captured individual timber rattlesnakes and surgically implanted intraperitoneal

transmitters (Holohil SI-2T) following procedures outlined by Reinert and Cundall (1982), using

machine-administered isoflurane to anesthetize animals via inhalation. Transmitter weight did

not exceed 5% of the snake's body weight, and in most instances was < 2%. We released snakes

at their original capture location within 24 hours of surgery and relocated individuals 2–3 times

per week from April through October 2016–2019. We tracked snakes between dawn and dusk

and regularly shifted the order in which individuals were tracked daily to avoid systematic bias,

though site access issues made it impractical to fully randomize when individuals were tracked.

Upon visually locating a snake, we recorded the location with a Global Positioning System (Garmin GPSmap 64s) to an estimated < 5m spatial accuracy.

Behavior Classification

We identified four behavioral and physiological states (hereafter referred to as "behaviors"): foraging, digestion, ecdysis, and gestation. We considered snakes to be foraging when they were observed in a stereotyped foraging posture (Reinert et al. 2011a, Tutterow et al. 2020). Our observations of snakes digesting a recent meal are limited by the difficulty in detecting a small bolus in a large-bodied snake, but we noted this whenever possible. These observations are, therefore, inherently biased toward snakes digesting larger, more obvious meals and may be prone to a high false negative detection rate. We noted observations of snakes in ecdysis as indicated by snakes having blue-gray eyes and dusky skin coloration. We confirmed observations of gestating females by sonogram shortly after snake emergence in April and May. We excluded all locations in which a snake was not visible and presumably underground or actively moving. When not moving and not clearly participating in the aforementioned behaviors, we designated snake behavior as resting. We excluded snake observations from analyses when there was uncertainty about behavioral classification.

Geospatial Habitat Features

We incorporated 15 geospatial habitat features, characterizing vegetation structure, plant species composition, and environmental variation, at varying resolutions (5-30 m), in our analysis of multiscale rattlesnake habitat use (Table 1). Digital canopy height (CHM) and elevation (DEM) models were developed from LiDAR data provided by the Ohio Geospatial Reference Program, collected in 2007 (http://ogrip.oit.ohio.gov/; accessed 13 October 2014). The LiDAR data featured two returns pulse^{-1} with an average spacing and density of 1.27 m and 0.27

returns m^{-2}, respectively. The CHM and DEM models incorporated conventional methods, using bilinear interpolation, at 5 m resolution. From the DEM, we developed layers on slope (°), solar radiation, and the Beer's transformed aspect (an index ranging 0–2, corresponding to southwestern to northeastern aspects, respectively). In addition to canopy height represented within the CHM, we developed three penetration ratios (number of returns <2 m in height divided by the number of returns <50 m and <10 m, including returns <2m; and the number of returns <1 m divided by the number of returns <5 m, including returns <1 m) to characterize overstory, midstory, and understory vegetation density at 30 m resolution (Adams and Matthews, 2019; Müller et al. 2010, Melin et al. 2018).

Woody plant composition was represented by three ordination axes within a gradient modeling approach developed from a separate study within the study area (Adams et al. 2019). Vegetation plot data, incorporating relative abundance profiles of trees and shrubs, including 99 woody plant taxa, were ordinated by non-metric multidimensional scaling (NMDS) and projected onto the landscape with terrain data and seasonal multispectral imagery provided by Landsat 8/OLI. Three subsequent floristic gradients synthesized moisture (NMDS1), successional (NMDS2), and elevational (NMDS3) variation among species responses within the vegetation data, at 30-m resolution. In addition, the plot data were used to develop geospatial layers on mean tree basal area (m^2 ha^{-1}) and density (stems ha^{-1}).

From April 2017 to December 2017, we collected temperature data across the property using HOBO Pendant data loggers (Model # UA-001-08) set to one-hour logging intervals. We placed 150 loggers randomly across our study site by staking each logger in place at ground level on the north side of a tree to limit the influence of canopy structure on thermal data. Following Fridley (2009) and George et al. (2015), we used these temperature data as a response variable in a linear mixed effects model with topographic and LiDAR-derived variables serving as predictive covariates. The final fitted model was used to predict the mean near-ground mid-summer temperature across our landscape. The purpose of this temperature variable was to capture relative differences in temperatures experienced by snakes between sites and, since temperature values were highly correlated across season and with snake body temperatures, we were able to use a single mid-summer date as an index value for these relative temperature differences. Finally, stand age was provided from the ODNR and a flowlines layer was transformed to a spatial grid (5 m) and used to summarize distances to the nearest stream (We extracted values for each covariate at all snake locations within different spatial scales of effect and centered and scaled all covariates prior to fitting statistical models. We assessed the relationship between these variables using a correlation matrix to ensure that all correlation coefficients were between -0.75 and 0.75.

Data Analysis

We modeled site use with generalized linear mixed effects Bernoulli models in a Bayesian framework using the brms package in R (Bürkner 2016; R Core Team 2020). For general habitat use models, we generated three non-use points for every use point by randomly shifting both the x and y coordinates of snake locations by up to $\pm$ 100 m. We fit a general habitat use model including observed snake locations and associated random points and considered snake site use as the response variable and the 15 environmental covariates as explanatory variables (Table 1). We ran each model with variables resampled at four representative spatial scales (5 m, 25 m, 55 m, and 105 m) using a moving window analysis (Miguet et al. 2016). We assessed the spatial scale at which each variable had the largest effect size and included these in a final multi-scale model. We also fit behavioral models to evaluate the relationship between habitat variables and the four focal behaviors (foraging, gestation, digestion, and ecdysis). For behavioral models, we used only observed snake locations. For each behavioral model, we coded snake locations according to the observed behavior and excluded all male snakes from the gravid snake model. All behavioral models included the 15 environmental variables used in the general site use model. We included snake identity as a random effect to account for individual variation between snakes. We ran models with four chains for 15,000 iterations with a warmup of 3,000 and no thinning. We visually inspected chains to confirm adequate mixing, and we confirmed convergence using the Gelman-Rubin statistic (all Rhat values=1). We used the Region of Practical Equivalence (ROPE) defined as $\pm0.1*SD$ of the response variable (Kruschke 2018) to guide our interpretation of the most influential covariates in each model and considered variables for which <11% of the posterior distribution was inside of ROPE to be particularly influential.

To generate multi-scale models, we ran four global, single-scale (5 m, 25 m, 55 m, and 105 m) models for a given response variable (general site use, gestation, ecdysis, digestion, or foraging). We then compared posterior distributions of parameters within each of these four models and selected the scale at which each variable had the largest estimated effect size for inclusion in the final, multi-scale model. We excluded any variables that had <95% of the posterior distribution with the same sign as the mean parameter estimate (probability of direction, PD).

We projected the probability of general site use, gestation, ecdysis, digestion, and foraging across the landscape using the posterior mean estimate from the most-supported behavioral models described above. To ensure we captured the most relevant parts of the landscape used by snakes, we buffered each observed snake location by 500 m and scaled each spatial layer by the mean and SD of the observed snake locations. We included each spatial covariate from Table 1, excluding only those parameters whose PD were < 95%. We then extracted the parts of the landscape that encompassed the top 20% most suitable habitat for each of the four behaviors and subsequently determined the percent overlap of each behavior on the landscape.

Results

We modeled the probability of site use by 43 timber rattlesnakes (19 males, 16 females, 8 juveniles) across four active seasons in relation to 15 environmental covariates (Table 2). Most snakes included in this analysis (30/43) were in our study for at least one full active season (mean days tracked = 512 ± 318 days, range = 6–1,020 days). Four covariates were strongly predictive of site use and ten other covariates were modestly predictive of site use (>11% of the posterior distribution inside of ROPE, but with a PD >95%) at one or more scales (Table 2). Solar radiation and slope (r =-0.68) and mean tree basal area and tree density (r = 0.74) were moderately correlated but were retained in our final models. We removed Beers transformed aspect from our final site use models because it was not predictive of site use at any scale. Overall, snakes were most likely to use warmer sites with greater solar radiation, greater mean tree basal area, but also increased disturbance as indicated by lower scores on NMDS 2 (Figure 1). Our single-scale model with all variables at the resolution of our original raster data (5 m) was the best supported model based on leave-one-out Information Criterion (LOOIC; Table 3).

Behavioral models fit the data well, and each behavior could be meaningfully predicted by a unique combination of variables assessed at different spatial scales (Table 4). A different suite of variables proved to be good predictors for each behavior, but no single variable or scale was useful for predicting all behaviors (Figures 2–5). Unlike the general habitat use model, observed snake behaviors were best predicted by variables assessed at multiple scales for all but foraging, which was best predicted when all variables were assessed at the 25-m scale (Table 3).

Our spatial projection of behaviors across the landscape found no more than 44% overlap between any two behaviors, with considerable variation between behaviors (Figure 6). Foraging was the most unique behavior (36.17% overlap with all other behaviors, collectively) and ecdysis was the least unique behavior (67.47% overlap with all other behaviors, collectively). Gestation and ecdysis had the most overlap of any two individual behaviors (44% overlap). By comparison, foraging and digestion had relatively little overlap (8.5% overlap).

Discussion

We found limited support for our hypothesis that snakes exhibiting more predictable patterns of behavior related to longer-term physiological states would be most affected by environmental variables at a larger spatial scale of effect. The best-supported model predicting our finest temporal scale behavior (foraging) only included variables at relatively fine spatial scales of effect (25 m). Conversely, the best-supported model predicting our coarsest temporal scale behavior (gestation) mostly included variables assessed at the 105 m scale. Importantly, mean tree basal area, tree density, and maximum canopy height were among the most important predictors of gestation and were included in the final model at our two coarsest scales. Collectively, low mean tree basal area, high tree density, and low maximum canopy height are indicative of early successional habitats and, given that these effects generally increased with scale, gestating females are clearly selecting for more disturbed macrohabitats. However, both digestion and ecdysis were also best predicted by variables assessed mostly at our coarsest scale

(105 m). Though these three behaviors do occur over longer temporal scales and were best predicted by covariates at larger spatial scales than foraging, it is important to also consider that foraging snakes are selecting habitat based on prey availability (Tutterow et al. 2020) and not necessarily on how suitable the habitat is for their own physiology.

Our behavioral models also shed light on the space use of timber rattlesnakes in a nuanced way that captures important habitat associations that traditional habitat use models might overlook. Our behavioral models indicate that within the generally warmer sites selected by snakes across the landscape, a temperature gradient exists that snakes are non-randomly exploiting based on their physiological state and behavior. Warmer mean temperatures were predictive of snakes that were gestating, undergoing ecdysis, or digesting a meal and the strength of this relationship increased with increasing spatial scale. These patterns indicate that snakes are moving to warmer parts of the landscape during these times, and not simply selecting warm microhabitats (e.g., canopy gaps). During ecdysis, gestation, and while digesting a meal, snakes maintain higher body temperatures (Gibson et al. 1989, Brown 1991), which was reflected in our behavioral models where temperature had a relatively strong effect on the probability of a snake exhibiting these behaviors. Conversely, our general site use model found a more modest effect of temperature on the probability of site use, likely because snakes are more likely to forage in relatively cool to moderate parts of the landscape and foraging locations accounted for 18% of our snake observations.

LiDAR-derived measures of canopy height and density were only moderately predictive of site use and site-specific behavior and generally only at finer scales. We suspect a disconnect exists in the scale at which these data are obtained and the scale at which vegetative structure influences snake habitat use. Plot-based, ground-collected data are typically used to describe the vegetative structure or microhabitat characteristics of sites used by snakes (Moore and Gillingham 2006, Sutton et al 2017) and remotely sensed data are likely less precisely capturing this fine-scale variability. Our best site use model also only included variables measured at the immediate snake location, further indicating that characteristics of the immediate environment (microhabitat) may be more important for timber rattlesnakes than variation in habitat at a larger scale (i.e., macrohabitat), as in other snake species (Steen et al. 2010, Bauder et al. 2018).

Stand age had no discernable effect on the probability of foraging or digesting a meal, but gestating females and snakes in ecdysis tended to use younger stands. While stand age had an increasingly positive effect on the probability of gestation with increasing spatial scale, the probability of ecdysis declined for stand age with increasing spatial scale. Both gestating snakes and snakes digesting meals were more likely to use sites with lower mean tree basal area, and this effect increased with increasing scale.

The forest successional metric NDMS 2 was also negatively predictive of gestation, which indicated that gravid snakes used disturbed sites with lower canopy height more frequently than non-gravid females. Snakes broadly selected for slightly more disturbed sites, synthesized among the vegetation data, suggesting this preference is even more pronounced in gravid females. Our results align with the known thermal requirements for gestating snakes (Charland and Gregory 1995, Reinert and Zappalorti 1988, Sprague and Bateman 2018). While snakes exhibiting thermoregulation-related behaviors generally selected warmer, more exposed sites, there are behavior-specific differences in the variables that influence their site use and the scale of effect for each of these variables. Ordination axis NMDS 1 had a relatively strong influence on the probability of foraging and indicates that snakes tended to forage at drier, oak-dominated sites relative to sites selected for other behaviors.

Our findings also emphasize the importance of multiple spatial scales for rattlesnake habitat use, especially for thermoregulation. Previous studies have also used multi-scale models to study the hierarchical nature of habitat selection in snakes (Moore and Gillingham 2006, Steen et al. 2010, Sutton et al 2017, Bauder et al. 2018), but ours is the first to present a temporal hierarchy of space use where scale of effect is dependent on behavior and synthesized simultaneously across a range of spatial scales. While fine-scale environmental data better predicted snake site use across the landscape and foraging behavior, multi-scale models were the best predictors of thermoregulation behavior. This indicates that resource selection by individual rattlesnakes is complex and influenced by environmental variation at different spatial scales.

To more precisely define potentially important habitats for timber rattlesnakes, we also spatially extrapolated our models across our study site to predict areas that are most suitable for each behavior. By calculating the overlap in site suitability between behaviors, we determined which behaviors were the most unique in their habitat associations. There was considerable overlap (26%) between our predicted most suitable regions for ecdysis and gestation, which is unsurprising given we often find gestating females in the same logs as snakes undergoing ecdysis (personal observation). There was considerably less overlap between other behaviors. The majority of regions (70%) identified as most suitable for foraging or digestion did not overlap our best predicted regions for either ecdysis or gestation.

Foraging is clearly a critical component of survival for rattlesnakes, but foraging snakes use habitat they would not normally use for thermoregulation. Without explicitly identifying these foraging locations, the importance of such habitats might be obscured in space use models. Our results highlight the importance of contextualizing models of space use for wildlife by explicitly accounting for both spatial and temporal variation in behavior. Traditional habitat selection and site use modeling weights the importance of a habitat based on its frequency of use

rather than how important the use of a site is to the survival or reproduction of an animal. Short-

term behaviors (such as foraging or digesting a meal) reliant on more subtle changes in local

habitats may be infrequent, relatively cryptic, and easily lost in a larger dataset of point locations.

We found that foraging rattlesnakes select for different habitat features and operate at a more

localized scale than snakes primarily concerned with thermoregulation. This not only

emphasizes the importance of a heterogeneous landscape, but also of interpreting the reasons an

animal uses a given site.

Although this approach may not be suitable for animals that are more difficult to directly

observe, we believe this is a good model for studying wild snakes and some other ectotherms

where direct observation is possible. By modeling predictors of site use, researchers can define

the general habitat preferences for a species and then situate predictive models of behavior, or

physiological state, within this broader context. This not only defines the preferred habitat types

for a species, but also pinpoints the more localized habitat features that may be important to

maintain within that preferred habitat type. Such models would ensure that managers are better

able to accommodate these specialized habitat needs and could lead to better outcomes for

threatened and endangered species.

Literature Cited

Adams, B.T. and S.N. Matthews. 2019. Diverse temperate forest bird assemblages demonstrate

closer correspondence to plant species composition than vegetation structure. Ecography

42, 1752–1764.

Adams, B.T., S.N. Matthews, M.P. Peters, A. Prasad, and L.R. Iverson. 2019. Mapping floristic gradients of forest composition using an ordination-regression approach with Landsat OLI and terrain data in the Central Hardwoods region. Forest Ecology and Management 434:87–98.

Bauder, J.M., D.R. Breininger, M.R. Bolt, M.L. Legare, C.L. Jenkins, B.B. Rothermel, and K. McGarigal. 2018. Multi-level, multi-scale habitat selection by a wide-ranging, federally threatened snake. Landscape ecology 33.5: 743–763.

Brown, W.S., D.W. Pyle, K.R. Greene, and J.B. Friedlaender. 1982. Movements and temperature relationships of timber rattlesnakes (*Crotalus horridus*) in northeastern New York. Journal of Herpetology 1982: 151–161.

Boyce, M.S., P.R. Vernier, S.E. Nielsen, and F.K. Schmiegelow. 2002. Evaluating resource selection functions. Ecological modelling 157:281–300.

Brown, W.S. 1991. Female reproductive ecology in a northern population of the timber rattlesnake, *Crotalus horridus*. Herpetologica:101–115.

Bürkner, P.C. 2016. brms: An R package for Bayesian multilevel models using Stan. Journal of Statistical Software 80:1–28.

Charland, M.B. and P.T. Gregory. 1995. Movements and habitat use in gravid and nongravid female garter snakes (Colubridae: Thamnophis). Journal of Zoology 236.4:543–561.

Clark, R. 2004. Timber Rattlesnakes (*Crotalus horridus*) use chemical cues to select ambush sites. Journal of Chemical Ecology 30.3:607–617.

Cleland, D.T., J.A. Freeouf, J.E. Keys, G.J. Nowacki, C.A. Carpenter, W.H. McNab. 2007. Ecological Subsections: Sections and Subsections for the Conterminous United States; Gen. Tech. Report WO-76D [Map on CD-ROM], Sloan, A.M. cartographer, presentation scale 1:3,500,000, colored; Department of Agriculture, Forest Service: Washington, DC, USA.

Cushman S.A. and K. McGarigal. 2004. Hierarchical analysis of forest bird species-environment relationships in the Oregon Coast Range. Ecological Applications 14:1090–1105,

Dalziel, B.D., J.M. Morales, and J.M. Fryxell. 2008. Fitting probability distributions to animal movement trajectories: using artificial neural networks to link distance, resources, and memory. The American Naturalist 172.2:248–258.

Ducey, L.J.R., 1982. Hardwood research in southern Ohio. Journal of Forestry, 80.8:488–489.

Ernst, C.H. and E.M. Ernst. 2003. Snakes of the United States and Canada. Smithsonian Books.

Fridley, J.D. 2009. Downscaling climate over complex terrain: high finescale (< 1000 m) spatial variation of near-ground temperatures in a montane forested landscape (Great Smoky Mountains). Journal of Applied Meteorology and Climatology 48.5:1033–1049.

Gardner-Santana, L.C. and S.J. Beaupre. 2009. Timber Rattlesnakes (*Crotalus horridus*) exhibit elevated and less variable body temperatures during pregnancy. Copeia 2:363–368.

George, A.D., F.R. Thompson III, and J. Faaborg. 2015. Using LiDAR and remote microclimate loggers to downscale near-surface air temperatures for site-level studies. Remote sensing letters 6.12:924–932.

George, A.D., G.M. Connette, F.R. Thompson III, and J. Faaborg. 2017. Resource selection by an ectothermic predator in a dynamic thermal landscape. Ecology and evolution 7.22:9557–9566.

Gibson, A.R., D.A. Smucny, and J. Kollar. 1989. The effects of feeding and ecdysis on temperature selection by young garter snakes in a simple thermal mosaic. Canadian Journal of Zoology, 67.1:19–23.

Greenwald, O.E. and M.E. Kanter. 1979. The effects of temperature and behavioral

thermoregulation on digestive efficiency and rate in corn snakes (*Elaphe guttata guttata*).

Physiological Zoology 52.3:398–408.

Hix, D.M. and J.N. Pearcy. 1997. Forest ecosystems of the Marietta Unit, Wayne National

Forest, southeastern Ohio: multifactor classification and analysis. Canadian Journal of

Forest Research 27:1117–1131.

Holland, J.D., D.G. Bert, and L. Fahrig. 2004. Determining the spatial scale of species' response

to habitat. Bioscience 54.3:227–233.

Huey, R.B. 1991. Physiological consequences of habitat selection. The American Naturalist

137:91–115.

Jackson, H.B. and L. Fahrig. 2012. What size is a biologically relevant landscape?. Landscape

Ecology 27.7:929–941.

Jackson N.D. and L. Fahrig. 2014. Landscape context affects genetic diversity at a much larger

spatial extent than population abundance. Ecology 95:871–881.

Jackson, H.B. and L. Fahrig. 2015. Are ecologists conducting research at the optimal scale?.

Global Ecology and Biogeography 24.1:52–63.

Kruschke, J.K. 2018. Rejecting or accepting parameter values in Bayesian estimation. Advances

in Methods and Practices in Psychological Science 1.2:270–280.

Lesmerises, R. and M.H. St-Laurent. 2017. Not accounting for interindividual variability can

mask habitat selection patterns: a case study on black bears. Oecologia 185.3:415–425.

Martin, W.H. 1993. Reproduction of the timber rattlesnake (*Crotalus horridus*) in the

Appalachian Mountains. Journal of Herpetology:133–143.

McGarigal, K., H.Y. Wan, K.A. Zeller, B.C. Timm, and S.A. Cushman. 2016. Multi-scale

habitat selection modeling: a review and outlook. Landscape ecology, 31.6:1161–1175.

Melin, M., S.A. Hinsley, R.K. Broughton, P. Bellamy, and R.A. Hill. 2018. Living on the edge:

utilizing lidar data to assess the importance of vegetation structure for avian diversity in

fragmented woodlands and their edges. Landscape Ecology 33:895–910.

Miguet, P., H.B. Jackson, N.D. Jackson, A.E. Martin, and L. Fahrig. 2016. What determines the

spatial extent of landscape effects on species? Landscape Ecology 31:1177–1194.

Moore, J.A. and J.C. Gillingham. 2006. Spatial ecology and multi-scale habitat selection by a

threatened rattlesnake: the eastern massasauga (*Sistrurus catenatus catenatus*). Copeia,

2006.4: 742–751.

Müller, J., J. Stadler, and R. Brandl. 2010. Composition versus physiognomy of vegetation as

predictors of bird assemblages: the role of lidar. Remote Sensing of Environment

114:490–495.

Neumann, W., S. Martinuzzi, A.B. Estes, A.M. Pidgeon, H. Dettki, G. Ericsson, and V.C.

Radeloff. 2015. Opportunities for the application of advanced remotely-sensed data in

ecological studies of terrestrial animal movement. Movement ecology 3.1:8.

Reinert, H.K. and D. Cundall. 1982. An improved surgical implantation method for radio-

tracking snakes. Copeia 3:702–705.

Reinert, H.K. and R.T. Zappalorti. 1988. Timber rattlesnakes (*Crotalus horridus*) of the Pine

Barrens: their movement patterns and habitat preference. Copeia:964–978.

Reinert, H.K., G.A. MacGregor, M. Esch, L.M. Bushar, and R.T. Zappalorti. 2011a. Foraging

ecology of Timber Rattlesnakes, *Crotalus horridus*. Copeia 3:430–442.

Semlitsch, R.D. 1979. The influence of temperature on ecdysis rates in snakes (genus *Natrix*)

    (Reptilia, Serpentes, Colubridae). Journal of Herpetology, 13.2:212–214.

Smith, A.C., L. Fahrig, and C.M. Francis. 2011. Landscape size affects the relative importance of

    habitat amount, habitat fragmentation, and matrix quality on forest birds. Ecography

    34.1:103–113.

Sprague, T.A. and H.L. Bateman. 2018. Influence of seasonality and gestation on habitat

    selection by northern Mexican gartersnakes (*Thamnophis eques megalops*). PloS one

    13.1.

Steen, D.A., J.M. Linehan, and L.L. Smith. 2010. Multiscale habitat selection and refuge use of

    common kingsnakes, *Lampropeltis getula*, in southwestern Georgia. Copeia 2010.2: 227–

    231.

Stevenson, R.D. 1985. The relative importance of behavioral and physiological adjustments

    controlling body temperature in terrestrial ectotherms. The American Naturalist

    126.3:362–386.

Stout, W. 1933. The charcoal iron industry of the Hanging Rock Iron District—its influence on

    the early development of the Ohio Valley. Ohio State Archaeol. Hist. Q. 42, 72–104.

Sutton, W.B., Y. Wang, C.J. Schweitzer, and C.J. McClure. 2017. Spatial ecology and multi-

    scale habitat selection of the Copperhead (*Agkistrodon contortrix*) in a managed forest

    landscape. Forest Ecology and Management 391: 469–481.

Thompson, C.M. and K. McGarigal. 2002. The influence of research scale on bald eagle habitat

    selection along the lower Hudson River, New York (USA). Landscape Ecology

    17.6:569–586.

Tutterow, A.M., A.S. Hoffman, J.L. Buffington, Z.T. Truelock, and W.E. Peterman. 2020. Prey-

    Driven Behavioral Habitat Use in a Low-Energy Ambush Predator. BioRxiv [**Preprint**].

    September 19, 2020 [cited 2021 Apr 7]. Available

    from: https://doi.org/10.1101/2020.09.17.301697.

Waldron, J.L., J.D Lanham, and S.H. Bennett. 2006. Using behaviorally-based seasons to

    investigate canebrake rattlesnake (*Crotalus horridus*) movement patterns and habitat

    selection. Herpetologica 62.4:389–398.

Zeller, K.A., K. McGarigal, P. Beier, S.A. Cushman, T.W. Vickers, and W.M. Boyce. 2014.

    Sensitivity of landscape resistance estimates based on point selection functions to scale

    and behavioral state: pumas as a case study. Landscape Ecology 29.3:541–557.

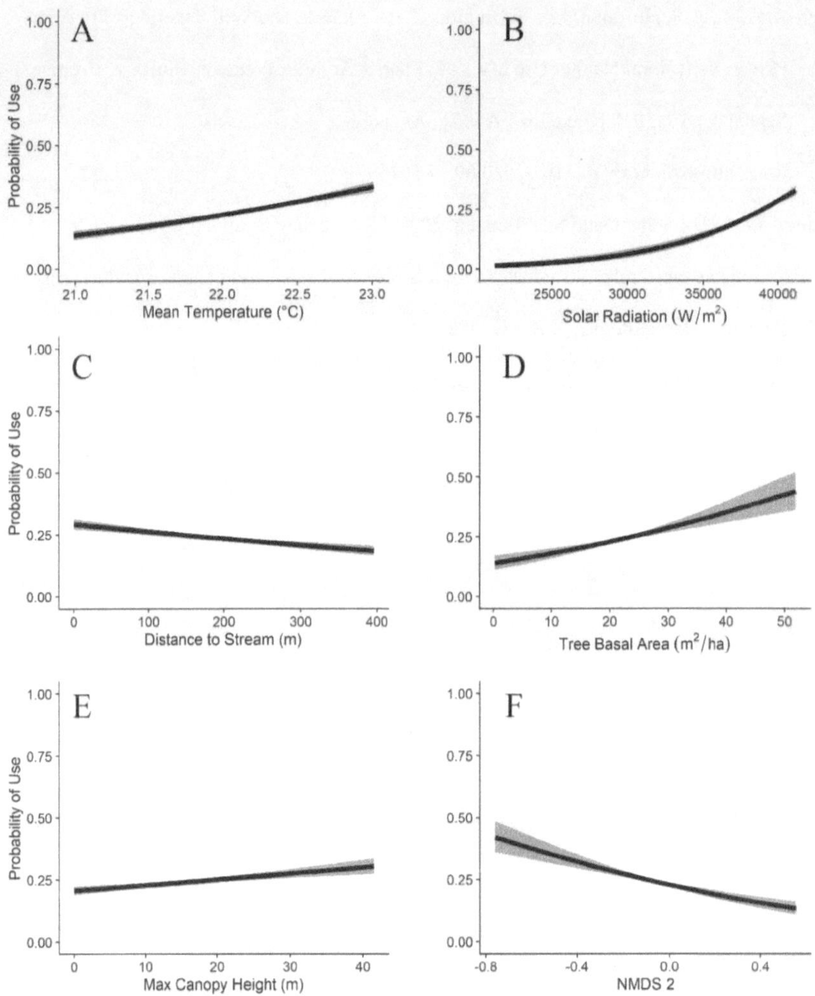

Figure 1.1. The effects of the six most influential environmental covariates (5-m scale) on the probability of general site use by timber rattlesnakes (*Crotalus horridus*) in southern Ohio. The probability of site use by snakes increases with increasing temperature (A), solar radiation (B), tree basal area (D), and maximum canopy height (E) and decreases with increasing distance to stream (C) and later forest successional stages (NMDS 2, F).

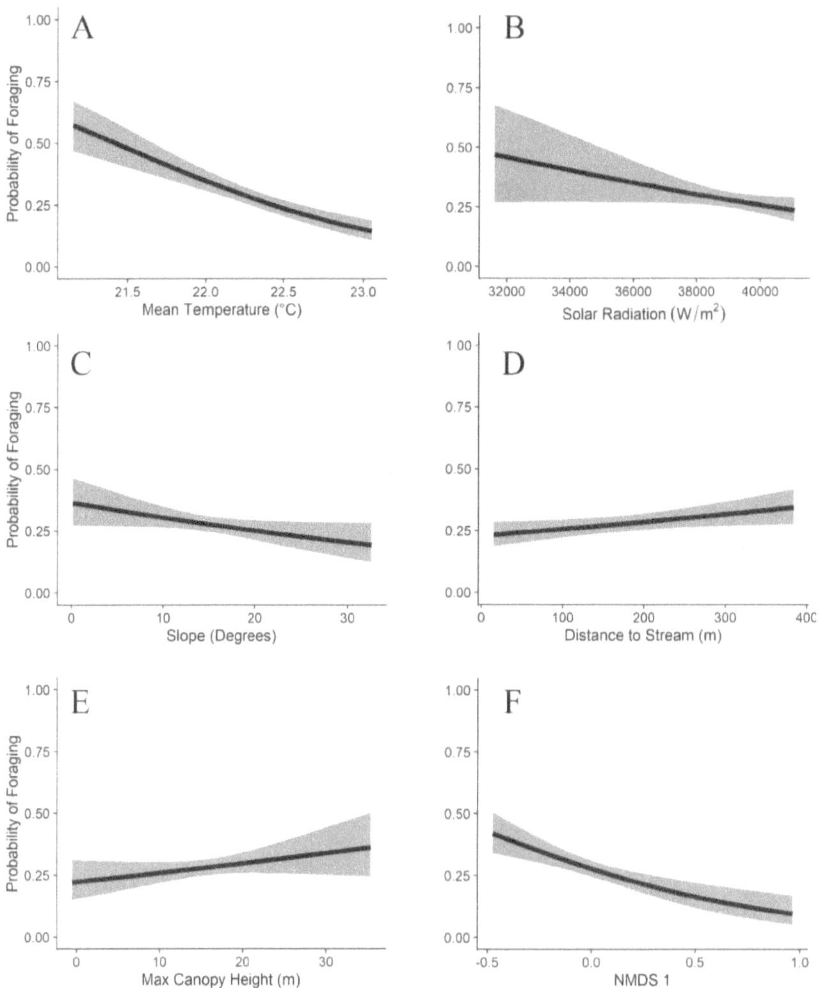

Figure 1.2. The effects of the six most influential covariates on the probability of timber rattlesnake (*Crotalus horridus*) foraging at a given site in southern Ohio. The probability of foraging decreases with increasing temperature (A), solar radiation (B), slope (C), and moisture (F) and increases with increasing distance to stream (D) and maximum canopy height (E).

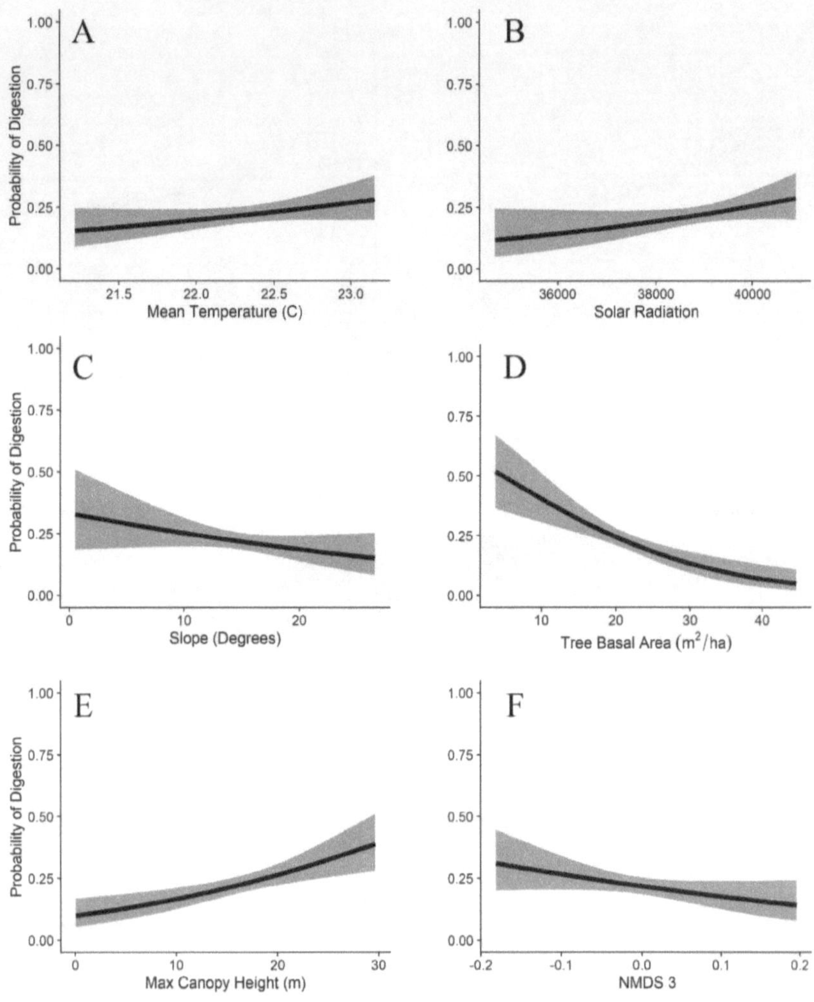

Figure 1.3. The effects of the six most influential covariates on the probability of timber rattlesnakes (*Crotalus horridus*) digesting a meal at a given site in southern Ohio. The probability of digestion increases with increasing temperature (A), solar radiation (B), and maximum canopy height (E) and decreases with increasing slope (C), tree basal area (D), and elevation (NMDS 3, F).

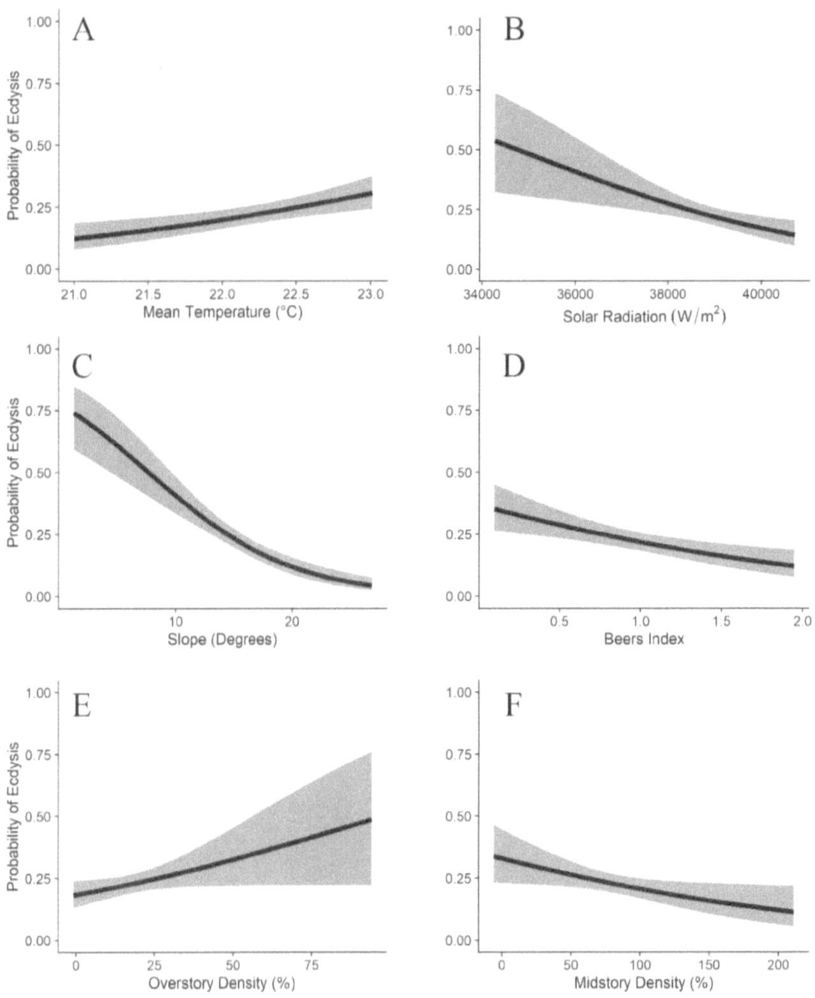

Figure 1.4. The effects of the six most influential covariates on the probability of timber rattlesnakes (*Crotalus horridus*) being in ecdysis at a given site in southern Ohio. The probability of ecdysis increases with increasing temperature (A) and overstory density (E) and decreases with increasing solar radiation (B), slope (C), values for Beers index (D), and midstory density (F).

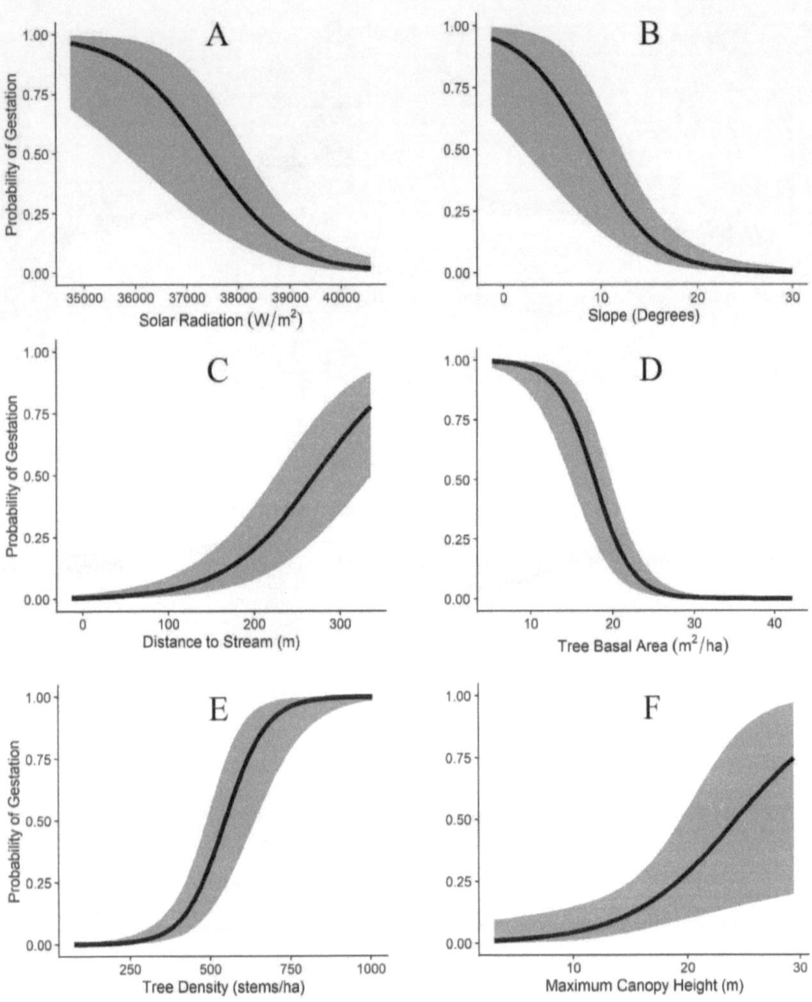

Figure 1.5. The effects of the six most influential covariates on the probability of female timber rattlesnakes (*Crotalus horridus*) gestating at a given site in southern Ohio. Probability of gestation decreases with increasing slope (A), solar radiation (B), and tree basal area (D) and increases with increasing distance to stream (C), tree density (E), and maximum canopy height (F).

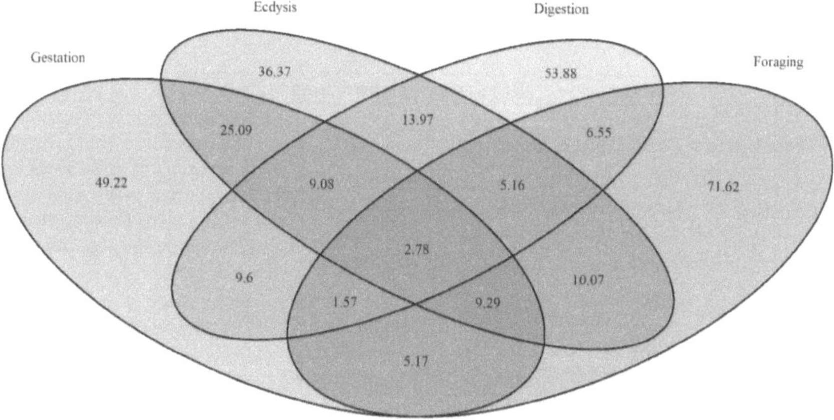

Figure 1.6. Venn diagram showing the spatial overlap in the most suitable (top 20%) habitat for timber rattlesnakes exhibiting four unique, time-varying behaviors in southern Ohio. Numbers represent the amount of land area (in hectares) in each category.

Table 1.1. List of geospatial habitat features used in modeling timber rattlesnake (*Crotalus horridus*) site use in southern Ohio. Covariate names are given alongside their data source, value range and units, and a summary of what each variable measures. Non-metric multidimensional scaled variables (NMDS) represent ordination axes.

Covariate	Source	Resolution	Value Range	Summary
Mean Temperature	Data Loggers	3 m	20–23° C	Mean ground temperature (measured on July 15) estimated from linear mixed effects model
Solar Radiation	DEM	3 m	29,708–41,129 WH/m^2	Cumulative solar radiation across active season (May to August)
Stand Age	Forestry	5 m	2–151 years	Year since plot was last clearcut as of 2019
Slope	DEM	3 m	0–41 degrees	Slope grade
Stream Distance	DEM	3 m	0–370 m	Distance to nearest stream
Beers	DEM	3 m	0–2	Beers transformed aspect
Tree Basal Area	Landsat	30 m	1–52 m^2 ha^{-1}	Mean tree basal area per hectare
Tree Density	Landsat	30 m	30–1,046 stems ha^{-1}	Mean stem density per hectare
Canopy Height	LiDAR	5 m	0–42 m	Maximum canopy height
Overstory	LiDAR	5 m	0–109	Proportion of LiDAR hits at canopy height
Midstory	LiDAR	5 m	0–210	Proportion of LiDAR hits at midstory height
Understory	LiDAR	5 m	32–385	Proportion of LiDAR hits at understory height
NMDS 1	Landsat	30 m	-0.50–1.03	Index related to moisture gradient; high values correspond to wetter soils and fewer oaks
NMDS 2	Landsat	30 m	-0.75–0.53	Index related to structure and forest succession; high values correspond to higher canopy height and less disturbance
NMDS 3	Landsat	30 m	-0.39–0.36	Index related to elevation; higher values correspond to higher elevations

Table 1.2. Estimated effects of environmental covariates on the probability of general site use by timber rattlesnakes (*Crotalus horridus*) in southern Ohio. Results represent our best-supported model with all covariates at their original 5-m resolution. The lower and upper 95% highest density intervals (HDI-low and HDI-high respectively) are presented alongside the probability of direction (PD), percent of the posterior distribution inside of the Region of Practical Equivalence (ROPE) using 89% of the posterior distribution. Bolded parameters represent covariates with < 1% of their posterior distribution inside ROPE.

Parameter	Estimate	HDI-low	HDI-high	PD	ROPE
Intercept	-1.15	-1.18	-1.11	1.00	0.00
Mean Temperature	**0.27**	**0.22**	**0.31**	**1.00**	**0.00**
Solar Radiation	**0.36**	**0.31**	**0.42**	**1.00**	**0.00**
Stand Age	-0.04	-0.08	0.00	0.95	1.00
Slope	0.11	0.06	0.16	1.00	1.00
Stream Distance	-0.15	-0.19	-0.10	1.00	0.97
Tree Basal Area	**0.24**	**0.17**	**0.31**	**1.00**	**0.05**
Tree Density	-0.07	-0.13	-0.01	0.98	1.00
Canopy Height	0.13	0.08	0.18	1.00	1.00
Overstory	0.01	-0.04	0.06	0.64	1.00
Midstory	-0.06	-0.11	-0.02	0.99	1.00
Understory	0.09	0.05	0.13	1.00	1.00
NMDS 1	0.12	0.07	0.16	1.00	1.00
NMDS 2	**-0.23**	**-0.28**	**-0.17**	**1.00**	**0.05**
NMDS 3	0.04	0.00	0.08	0.97	1.00

Table 1.3. Model comparisons using leave one out information criteria (LOOIC) for timber rattlesnake site use (Use) and behavior in southern Ohio. For behavioral models, model name prefixes correspond to particular behaviors (For = foraging, Dig = digestion, Ecd = ecdysis, Ges = gestation) and all model names end with an indication of the scale at which variables were smoothed (5 m, 25 m, 55 m, 105 m, MS = multi-scale).

Model	K	LOOIC	ΔLOOIC	w
Use_5	14	14485.780	0.0000	0.9837
Use_25	14	14493.990	-8.2012	0.0163
Use_MS	14	14507.060	-21.2710	0.0000
Use_55	14	14553.550	-67.7608	0.0000
Use_105	14	14664.840	179.0595	0.0000
For_25	7	2385.960	0.0000	0.7460
For_5	7	2388.816	-2.8561	0.1789
For_MS	7	2391.541	-5.5805	0.0458
For_55	7	2392.488	-6.5276	0.0285
For_105	7	2399.574	-13.6133	0.0008
Dig_MS	6	785.212	0.0000	0.4679
Dig_105	6	785.239	-0.0271	0.4616
Dig_55	6	789.039	-3.8273	0.0690
Dig_25	6	796.768	-11.5562	0.0014
Dig_5	6	802.412	-17.2000	0.0001
Ecd_MS	13	2230.439	0.0000	1.0000
Ecd_105	13	2252.125	-21.6862	0.0000
Ecd_5	13	2264.357	-33.9181	0.0000
Ecd_55	13	2265.355	-34.9162	0.0000
Ecd_25	13	2271.700	-41.2613	0.0000
Ges_MS	14	673.375	0.0000	0.9891
Ges_5	14	683.946	-10.5707	0.0050
Ges_25	14	684.047	-10.6715	0.0048
Ges_55	14	686.966	-13.5909	0.0011
Ges_105	14	726.450	-53.0745	0.0000

Table 1.4. Model-specific effects of each covariate on the probability of timber rattlesnakes (*Crotalus horridus*) exhibiting one of four distinct behaviors. Parameter estimates are given alongside 89% highest density intervals and the scale at which each variable had the greatest effect. All listed variables have strong support for having the estimated effect on behavior (probability of direction ≥95%). Bolded text indicates that greater than 89% of the posterior distribution fell outside of the ROPE for that variable. Results presented are from reduced, multi-scale models for digestion, ecdysis and gestation and for a reduced, single-scale model (25-m) for foraging. The identified scale (5–105 m) for each covariate is indicated below the parameter estimate.

Parameter	Foraging	Digestion	Ecdysis	Gestation
Mean Temperature	**-0.44 (-0.56 to -0.32)** 25 m	0.17 (-0.01 to 0.35) 5 m	0.25 (0.13 to 0.37) 5 m	**0.94 (0.55 to 1.33)** 105 m
Solar Radiation	-0.17 (-0.32 to -0.03) 25 m	0.21 (0.01 to 0.42) 105 m	**-0.35 (-0.54 to -0.17)** 105 m	**-1.50 (-2.06 to -0.92)** 105 m
Stand Age			-0.16 (-0.27 to -0.04) 25 m	-0.07 (-0.42 to -0.29) 105 m
Slope	-0.16 (-0.29 to -0.03) 25 m	-0.19 (-0.41 to 0.02) 55 m	**-0.70 (-0.87 to -0.54)** 105 m	**-1.25 (-1.72 to -0.81)** 105 m
Stream Distance	0.14 (0.03 to 0.24) 25 m		-0.19 (-0.32 to -0.06) 105 m	**1.79 (1.41 to 2.20)** 25 m
Beers	0.11 (-0.01 to 0.24) 25 m		**-0.33 (-0.48 to -0.16)** 105 m	**-0.74 (-1.18 to -0.32)** 105 m
Tree Basal Area		**-0.52 (-0.73 to -0.32)** 105 m		**-3.31 (-4.00 to -2.60)** 55 m
Tree Density			-0.24 (-0.35 to -0.13) 105 m	**1.92 (1.40 t0 2.45)** 105 m
Canopy Height	0.16 (-0.02 to 0.36) 25 m	0.42 (0.22 to 0.63) 105 m	-0.16 (-0.52 to 0.18) 105 m	**1.51 (0.56 to 2.47)** 105 m
Overstory			0.29 (0.05 to 0.53) 105 m	**0.53 (0.00 to 1.04)** 25 m
Midstory			-0.28 (-0.48 to -0.08) 5 m	0.16 (-0.41 to 0.76) 25 m
Understory			0.15 (0.03 to 0.29) 5 m	**1.20 (0.80 to 1.59)** 105 m
NMDS 1	**-0.34 (-0.48 to -0.21)** 25 m		-0.18 (-0.35 to -0.01) 55 m	**0.47 (0.06 to 0.92)** 105 m
NMDS 2				**-0.70 (-1.21 to -0.18)** 5 m
NMDS 3		-0.18 (-0.36 to -0.01) 105 m	0.09 (-0.01 to 0.20) 5 m	

Chapter 2: Using forest structure and composition metrics to predict timber rattlesnake (*Crotalus horridus*) site use

Introduction

Modern forest management practices are focused on affecting ecosystem-scale changes by monitoring and subsequently altering tree communities. In eastern North America, this often means restoring oak-dominated forests and combatting forest mesophication through the use of prescribed fire and selective timber harvest (Nowacki and Abrams 2008, Arthur et al. 2015). A shift away from oak-dominated forests is undesirable because oaks are economically valuable for the timber industry, promote biodiversity, and produce the most ecologically important type of hard mast food resource for wildlife (McShea and Healy 2002, Fralish, 2004, Smith 2006, Brose et al. 2008). In order to maintain or promote oak dominance, land managers use forest stand structure metrics (e.g. basal area, relative density) to monitor oak regrowth and stocking and prioritize management of sites where oak regrowth will be most effective (Brose et al. 2008, Iverson et al. 2019).

The Silviculture of Alleghany Hardwoods (SILVAH; Marquis 1992) program introduced a means by which to standardize measurement of forest structure, growth, and development and provide silvicultural prescriptions for forests across much of eastern North America (Stout and Brose 2014). This has since been adapted to more explicitly generate silvicultural prescriptions for oak regeneration (Iverson et al. 2018, Brose 2019). Though some metrics used in this program, such as basal area and tree density, are commonly used by wildlife researchers to describe wildlife habitat use (Payer and Harrison 2003, Hyslop 2007), fewer wildlife studies more comprehensively incorporate this framework in modeling spatial ecology of forest species (Adams et al. 2019, Adams and Matthews 2019).

Throughout many eastern deciduous forests, timber rattlesnakes (*Crotalus horridus*) are a species that foresters are asked to consider in their management plans (IDFW 2015, IDNR 2015, ODW 2015). Though not federally protected, timber rattlesnakes have declined throughout their range due to persecution and habitat loss and are considered a protected species in multiple states (Ulev 2008). Previous studies have found little effect of forest management activities on timber rattlesnakes (Reinert et al. 2011b, MacGowan et al. 2017), but most studies on this species have focused on describing the extent of their movements and the importance of microhabitat features on their habitat use (Reinert and Zappalorti 1988, Adams 2005, Ashley 2020). No attempts have yet been made to holistically incorporate well-studied metrics of forest structure and composition to better understand habitat use in this forest-reliant species.

In order to more directly link the management of forest ecosystems in the eastern United States to the habitat needs of timber rattlesnakes, we paired traditional point-based snake telemetry data with both variable-radius prism plot data and non-plot based micro-habitat data, allowing us to estimate tree diversity, size, and density at sites used by rattlesnakes. Although measuring understory tree regeneration is an important component of SILVAH prescriptions (Marquis 1992), we did not include this in our macrohabitat sampling because we believed our microhabitat measures would capture understory vegetative density at a scale that would be more relevant to snakes. Our behavior-mediated habitat selection models gave some indication that forest structure was important for timber rattlesnakes so we hypothesized that this structure might also be linked to forest composition (see Chapter 1 or Hoffman et al. 2020). Given that gestating and shedding timber rattlesnakes selected for the kind of dry, open, and disturbed sites where oaks often thrive, we predicted that snakes might also specifically select for oak-dominated sites.

Methods

Study Site

Vinton Furnace State Forest (VFSF) is a 4,892-ha property located in the heavily dissected unglaciated hills of southeastern Ohio. The region is comprised largely of second growth, mixed-mesophytic forests recovering from widespread exploitation during the 1800's (Stout 1933). Oaks (*Quercus* spp.) are the dominant tree species, especially on ridges and southwest slopes, while sugar maple (*Acer saccharum*), American beech (*Fagus grandifolia*), and yellow poplar (*Liriodendron tulipifera*) are more prevalent on the mesophytic northeast slopes and bottomlands (Hix and Pearcy 1997; Adams et al. 2019). There are active and ongoing forest management activities at VFSF to promote habitat heterogeneity, including even- and uneven-aged silvicultural prescriptions and prescribed fire (Ducey 1982).

Data Collection

We collected data on forest structure and tree species composition across a randomly stratified sample of points from a database including 43 timber rattlesnakes tracked over the course of four years (2016–2019), and at an equal number of paired random points that ranged from 50 to 150 meters from each snake point. We stratified selection by behavior (foraging), condition (gravid or in ecdysis), and individual, to capture variability known to influence site selection (Hoffman et al. 2020). We used a variable-radius plot with a 10-factor prism and recorded tree species and DBH for each tree within the selected plots.

We collected micro-habitat data during 2016 and 2017 at locations used by 23 radio-telemetered timber rattlesnakes and at associated random points. Each random point was located at a random distance (1–30 m) and random bearing (0 – 360°) from its associated snake location. All snake locations had at least one paired random point and we collected data at three associated random points for locations at which snakes were observed foraging, in ecdysis, or gravid. Upon locating a snake, we flagged the location and returned to collect micro-habitat data after the snake had left to avoid unnecessary disturbance. We used two meter-long PVC poles to visually delineate a 1 m² plot around each location when collecting data.

Variables

We calculated five site-level covariates from the variable radius plot data related to forest composition and structure including total DBH for all trees within the plot, tree density (number of trees/ha), relative tree density (Stout et al. 1987), a white oak importance score calculated as the sum of both the relative dominance of white oak (total white oak basal area/total tree basal area) and the relative density of white oak (number of white oak/number of trees), and Shannon's Diversity Index to estimate tree diversity. We also included ecological land type phases (ELTP) to classify each site on a topographic vegetational gradient. ELTP is derived from a digital elevation model and broken down into topographic categories based largely off of four commonly used indices (Beer's Index, topographic position index, and slope percent; Iverson et al. 2019).

We calculated eight site-level covariates from the micro-habitat data collected from 2016–2017. Within a 1-m² plot centered on each location, we estimated the percent coverage of herbaceous vegetation (< 1 m tall), downed woody debris, exposed rock, and tree trunks. We also measured the distance from the plot's center point to the nearest overstory tree (DBH > 7 cm), understory tree (DBH < 7 cm), and fallen log. Finally, we used densiometers (during 2016) and smart phone cameras with attached fisheye lenses (during 2017) to measure canopy cover at each location, facing due north. We used the image processing package 'EBImage' (Pau et al., 2010) to analyze all smart phone canopy cover images. This program assigned pixel intensity values based on the average difference in the gray level of the canopy vegetation and the sky background to estimate percent canopy cover.

Data Analysis

We evaluated site use by rattlesnakes using generalized linear mixed effects Bernoulli models fit in a Bayesian framework using the brms package in R (Bürkner 2016; R Core Team 2020). We fit two models including observed snake locations and associated random points and considered snake site use as the response variable for both models. We used six forest structure and composition covariates as explanatory variables in one model and eight microhabitat covariates as explanatory variables in our other model. We included snake (individual), sex (male or female), behavior (foraging or not), and condition (gravid/in ecdysis or not) as random effects to account for individual variation between snakes and fine-scale behavioral-state variation in both models. We ran models with four chains for 15,000 iterations with a warmup of 3,000 and no thinning. We visually inspected chains to confirm adequate mixing, and we confirmed convergence using the Gelman-Rubin statistic (all Rhat values≤1). We used an 89% Region of Practical Equivalence (ROPE) defined as ±0.1*SD of the response variable (Kruschke 2018) to guide our interpretation of the most influential covariates in each model and considered variables for which <15% of the posterior distribution was inside of ROPE to be particularly influential.

Results

We collected forest structure and composition data from a total of 356 plots representing 178 sites used by 43 radio-telemetered snakes (19 males, 16 females, 8 juveniles) and 178 paired random points. We modeled the probability of snake site use across this sample using six forest structure and composition measures as predictor variables. Of the six covariates related to vegetation and/or topography we included in the model, only two were influential (<15% of posterior distribution inside ROPE) in predicting snake site use (Table 2.1). There was a moderate probability of snakes using sites with low relative tree density and a very low probability of snakes using sites with high relative tree density (Figure 2.1). Snakes were also more likely to use drier, more upland land cover types over mesic bottomlands (Figure 2.2). Metrics related to tree basal area and DBH, tree species diversity, and oak dominance were not informative in our model.

We collected micro-habitat data from a total of 775 points representing 337 sites used by 23 radio-telemetered snakes (10 males, 9 females, 4 juveniles) and 438 associated random points. We modeled the probability of snake site use across this sample using eight micro-habitat metrics as predictor variables. Of these eight covariates, only three were influential (<15% of posterior distribution inside ROPE) in predicting snake site use (Table 2.2). Snakes were more likely to use sites with a higher percentage of herbaceous ground cover (Figure 2.3) and downed woody debris in the form of logs and fallen limbs (Figure 2.4) and were more likely to use sites that were closer to fallen logs (Figure 2.5).

Discussion

Relative tree density and ELTP were useful in predicting timber rattlesnake site use in southern Ohio. A relative tree density metric first developed by Stout (1987), which accounts for variation in growth structure between different tree species, was consistently predictive of snake site use, though its impact was modest (Figure 2.1). Higher forest densities appear to have a negative effect on the probability that rattlesnakes will use a site, indicating an aversion to overly dense forest habitat. Though this metric is not a direct correlate of canopy openness or incoming solar radiation, more densely stocked stands will generally experience less light penetration through the canopy layers (Brose et al. 2008). Likewise, canopy cover was not an important predictor of site use in our micro-habitat model, but snakes were more likely to use sites with more herbaceous ground cover. Denser and more diverse undergrowth is often associated with canopy gaps in otherwise closed-canopy forests (Goldblum 1997). These findings are not surprising given that temperature plays a critical role in the efficiency of prey digestion and metabolization for snakes (Beaupre and Zaidan 2012) and that timber rattlesnakes generally seek out and bask in more open, disturbed sites, except when foraging (Hoffman et al. 2020).

The ELTP land classification system was the single most predictive variable in our forest structure and composition model, indicating timber rattlesnakes preferably use the more exposed ridges, and upper/southwest slopes over bottomlands. This land cover classification system is a critical component of oak restoration efforts because it describes the landscape largely along a topographic-moisture gradient that closely matches floristic gradients and communities (Iverson et al. 2018). Our model indicates that the land cover types rattlesnakes prefer are also likely to be sites that are high priority sites for oak restoration. Given that white oak dominance did not affect site use, it is likely that this trend reflects an overlap in preferred habitat between oaks and rattlesnakes. However, we believe that the combined importance of more topographically exposed habitats and less dense tree communities is indicative of a preference for forests dominated by more open-growing tree species, including most oaks.

Snakes were also much more likely to use sites with more downed woody debris, and sites nearer to fallen logs. Downed logs are frequently used by rattlesnakes to ambush prey (Reinert et al. 2011a, Tutterow et al. 2020), and by gravid females during gestation (Reinert and Zappalorti 1988), but our results indicate that snakes more broadly select sites with more downed woody debris cover. At our study site, snakes sheltered in or under logs throughout the year, but especially during cold weather or when the surrounding landscape was relatively open (i.e., little vegetative cover). Therefore, we believe it will generally benefit timber rattlesnakes to leave woody debris on the ground following timber harvest, and to keep log piles further from road edges to minimize road mortality.

Our results highlight two broadly important considerations for forest managers in areas where timber rattlesnakes are present. Timber rattlesnakes are more likely to use and be present at upland sites that would be conducive to oak regeneration efforts and may therefore experience more timber harvesting and prescribed fire. However, these management activities are ultimately contributing to maintaining the more open sites that rattlesnakes prefer (Iverson et al. 2017). Although snakes using these areas, could experience increased mortality rates from management activities, Reinert et al. (2011b) did not observe increased mortality in a Pennsylvania rattlesnake population during timber harvesting operations. He attributed this largely to his communication with foresters which ultimately increased their awareness and avoidance of rattlesnakes, suggesting that promoting rattlesnake awareness may be key to mitigating mortality associated with forest management activity. Likewise, the mortality observed by Beaupre and Douglas

(2012) following a prescribed fire could be largely attributed to a lack of communication between managers and researchers.

Our findings highlight that timber rattlesnake habitat use is best predicted by a relatively small subset of commonly used forest structure metrics. We believe that forest management aimed at reducing the effects of mesophication is likely to benefit timber rattlesnakes, but such treatments should be conducted under the guidance of, or in communication with, wildlife biologists and researchers that can help lower the mortality risk for resident rattlesnakes.

Literature Cited

Adams, B.T. and S.N. Matthews. 2019. Diverse temperate forest bird assemblages demonstrate closer correspondence to plant species composition than vegetation structure. Ecography, 42.10: 1752–1764.

Adams, B.T., S.N. Matthews, M.P. Peters, A. Prasad, and L.R. Iverson. 2019. Mapping floristic gradients of forest composition using an ordination-regression approach with Landsat OLI and terrain data in the Central Hardwoods region. Forest Ecology and Management 434:87–98.

Adams, J.P. 2005. Home range and behavior of the timber rattlesnake (*Crotalus horridus*). Huntington, WV, Marshall University. book.

Arthur, M.A., B.A. Blankenship, A.L. Schörgendorfer, L. David, H.D. Alexander. 2015. Changes in stand structure and tree vigor with repeated prescribed fire in an Appalachian hardwood forest. Forest Ecology and Management. 340: 46–61. 16 p. 10.1016/j.foreco.2014.12.025.

Ashley, J.M. 2020. Habitat Selection of a Timber Rattlesnake (*Crotalus horridus*) population in Middle Tennessee. Murfreesboro, TN, Middle Tennessee University. book.

Beaupre, S.J. and L.E. Douglas. 2012. Responses of timber rattlesnakes to fire: lessons from two prescribed burns. In: Dey, D.C., M.C. Stambaugh, S.L. Clark, and C.J. Schweitzer (eds.) Proceedings of the 4th fire in eastern oak forests conference; 2011 May 17–19; Springfield, MO. Gen. Tech. Rep. NRS-P-102. Newtown Square, PA: US Department of Agriculture, Forest Service, Northern Research Station: 192-204: 192–204.

Beaupre, S.J. and F. Zaidan. 2012. Digestive performance in the timber rattlesnake (*Crotalus horridus*) with reference to temperature dependence and bioenergetic cost of growth. Journal of Herpetology 46.4: 637–642.

Brose, P.H., K.W. Gottschalk, S.B. Horsley, P.D. Knopp, J.N. Kochenderfer, B.J. McGuinness, G.W. Miller, T.E. Ristau, S.H. Stoleson, and S.L. Stout. 2008. Prescribing regeneration treatments for mixed-oak forests in the Mid-Atlantic region. Gen. Tech. Rep. NRS-33. Newtown Square, PA: U.S. Department of Agriculture, Forest Service, Northern Research Station: 100 https://doi.org/10.2737/NRS-GTR-33.

Brose, P.H., 2019. Expanding the SILVAH decision support system to be applicable to the mixed-oak forests of the mid-Atlantic region. *In* SILVAH: 50 years of science-management cooperation (ed. S.L. Stout). Proceedings of the Allegheny Society of American Foresters training session; 2017 Sept. 20–22; Clarion, PA. Gen. Tech. Rep. NRS-P-186.

Bürkner, P.C. 2016. brms: An R package for Bayesian multilevel models using Stan. Journal of Statistical Software 80:1–28.

Ducey, L.J.R., 1982. Hardwood research in southern Ohio. Journal of Forestry, 80.8:488–489.

Ernst, C.H. and E.M. Ernst. 2003. Snakes of the United States and Canada. Smithsonian Books.

Fralish, J.S. 2004. The keystone role of oak and hickory in the central hardwood forest. In:
Spetich, M.A., ed. Upland oak ecology symposium: history, current conditions, and
sustainability. Gen. Tech. Rep. SRS-73. Asheville, NC: U.S. Department of Agriculture,
Forest Service, Southern Research Station: 78–87.

Goldblum, D. 1997. The effects of treefall gaps on understory vegetation in New York
State. Journal of Vegetation Science 8.1:125–132.

Hix, D.M. and J.N. Pearcy. 1997. Forest ecosystems of the Marietta Unit, Wayne National
Forest, southeastern Ohio: multifactor classification and analysis. Canadian Journal of
Forest Research 27:1117–1131.

Hoffman, A.S., A.M. Tutterow, M.R. Gade, B.T. Adams, and W.E. Peterman. 2020. Variation in
behavior drives multiscale responses to habitat conditions in timber rattlesnakes (Crotalus
horridus). BioRxiv [**Preprint**]. December 4, 2020 [cited 2020 Dec 31]. Available
from: https://doi.org/10.1101/2020.12.03.410290.

Hyslop, N.L., 2007. Movements, habitat use, and survival of the threatened eastern indigo snake
(*Drymarchon couperi*) in Georgia. Doctoral book, University of Georgia.

Illinois Department of Natural Resources (IDNR). 2015. Illinois state wildlife action plan species
of greatest conservation need. Springfield, IL, USA.

Indiana Division of Fish and Wildlife (IDFW). 2015. Indiana state wildlife action plan species of
greatest conservation need. Indianapolis, IN, USA.

Iverson, L.R., T.F. Hutchinson, M.P. Peters, and D.A. Yaussy. 2017. Long-term response of oak-
hickory regeneration to partial harvest and repeated fires: influence of light and moisture.
Ecosphere. 8.1: e01642. 24 p. http://dx.doi.org/10.1002/ecs2.1642.

Iverson, L.R., M.P. Peters, J.L. Bartig, J. Rebbeck, T.F. Hutchinson, S.N. Matthews, S. Stout.

2018. Spatial modeling and inventories for prioritizing investment into oak-hickory

restoration. Forest Ecology and Management 424: 355–366.

https://doi.org/10.1016/j.foreco.2018.05.018.

Iverson, L.R., J.L. Bartig, G.J. Nowacki, M.P. Peters, J.M. Dyer, T.F. Hutchinson, S.N.

Matthews, and B.T. Adams. 2019. USDA Forest Service section, subsection, and

landtype descriptions for southeastern Ohio. Research Map NRS-10. Newtown Square,

PA: U.S. Department of Agriculture, Forest Service, Northern Research Station. 68 p.

https://doi.org/10.2737/NRS-RMAP-10.

Kruschke, J.K. 2018. Rejecting or accepting parameter values in Bayesian estimation. Advances

in Methods and Practices in Psychological Science 1.2:270–280.

MacGowan, B.J., A.F. Currylow, and J.E. MacNeil. 2017. Short-term responses of Timber

Rattlesnakes (*Crotalus horridus*) to even-aged timber harvests in Indiana. Forest Ecology

and Management, 387: 30–36.

Marquis, D.A. 1992. User's guide to SILVAH: Stand analysis, prescription, and management

simulator program for hardwood stands of the Alleghenies 162. US Department of

Agriculture, Northeastern Forest Experiment Station.

McShea, W.J. and W.M. Healy (eds.). 2002. Oak forest ecosystems: ecology and management

for wildlife. Baltimore, MD: Johns Hopkins University Press: 432.

Ohio Division of Wildlife (ODW). 2015. Ohio's State Wildlife Action Plan. Columbus, Ohio,

USA.

Nowacki, G.J. and M.D. Abrams. 2008. The demise of fire and "mesophication" of forests in the

eastern United States. BioScience, 58.2: 123–138.

Payer, D.C. and D.J. Harrison. 2003. Influence of forest structure on habitat use by American marten in an industrial forest. Forest ecology and Management 179.1–3: 145–156.

Pau G., F. Fuchs, O. Sklyar, M. Boutros, W. Huber. 2010. EBImage—an R package for image processing with applications to cellular phenotypes. Bioinformatics 26:979–981.

Reinert, H.K. and R.T. Zappalorti. 1988. Timber Rattlesnakes (*Crotalus horridus*) of the Pine Barrens: Their movement patterns and habitat preference. Copeia 4:964–978.

Reinert, H.K., G.A. MacGregor, M. Esch, L.M. Bushar, and R.T. Zappalorti. 2011a. Foraging ecology of Timber Rattlesnakes, *Crotalus horridus*. Copeia 3:430–442.

Reinert, H.K., W.F. Munroe, C.E. Brennan, M.N. Rach, S. Pelesky, and L.M. Bushar. 2011b. Response of Timber Rattlesnakes to commercial logging operations. Journal of Wildlife Management 75.1:19–29.

Smith, D.W. 2006. Why sustain oak forests? In: Dickinson, M.B., ed. Fire in eastern oak forests: delivering science to land managers. Gen. Tech. Rep. NRS-P-1. Newtown Square, PA: U.S. Department of Agriculture, Forest Service, Northern Research Station: 62–71.

Stout, W. 1933. The charcoal iron industry of the Hanging Rock Iron District—its influence on the early development of the Ohio Valley. Ohio State Archaeol. Hist. Q. 42: 72–104.

Stout, S.L., D.A. Marquis, R.L. Ernst. 1987. A relative density measure for mixed species stands. Journal of Forestry 85.6: 45–47.

Stout, S.L. and P.H. Brose. 2014. The SILVAH saga: 40+ years of collaborative hardwood research and management highlight silviculture. Journal of Forestry 112.5:434–439.

Tutterow, A.M., A.S. Hoffman, J.L. Buffington, Z.T. Truelock, and W.E. Peterman. 2020. Prey-Driven Behavioral Habitat Use in a Low-Energy Ambush Predator. BioRxiv [**Preprint**].

September 19, 2020 [cited 2021 Apr 7]. Available

from: https://doi.org/10.1101/2020.09.17.301697.

Ulev, E. 2008. *Crotalus horridus*. In: Fire Effects Information System, [Online]. U.S.

Department of Agriculture, Forest Service, Rocky Mountain Research Station, Fire

Sciences Laboratory (Producer). Available:

www.fs.fed.us/database/feis/animals/reptile/crho/all.html [2021, January 24].

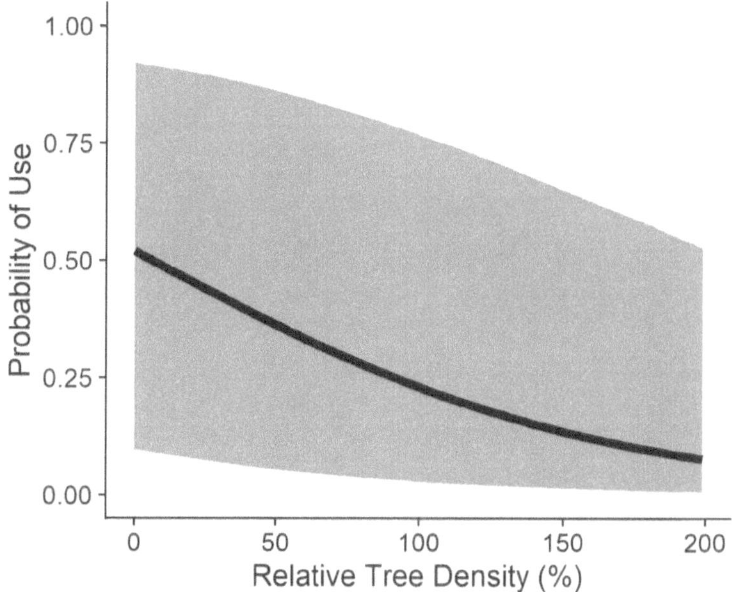

Figure 2.1. The effect of relative tree density on the probability of site use by timber rattlesnakes in southern Ohio. Timber rattlesnakes were more likely to use sites with lower relative tree density. Shaded areas represent 95% credible intervals.

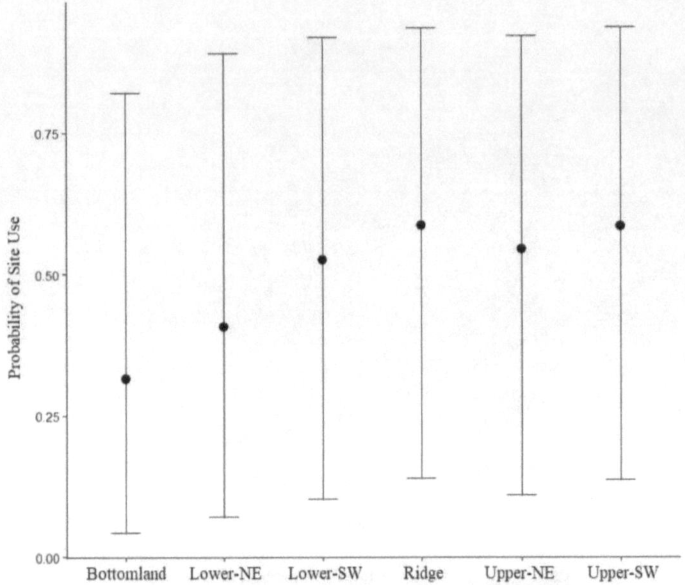

Figure 2.2. The effect of Ecological Land Type Phases (ELTP) on the probability of site use by timber rattlesnakes in southern Ohio. ELTP categories included bottomland, lower northeast slope (Lower-NE), lower southwest slope (Lower-SW), ridge, upper northeast slope (Upper-NE), and upper southwest slope (Upper-SW). Timber rattlesnakes were more likely to use upper slope and ridge sites than bottomland sites.

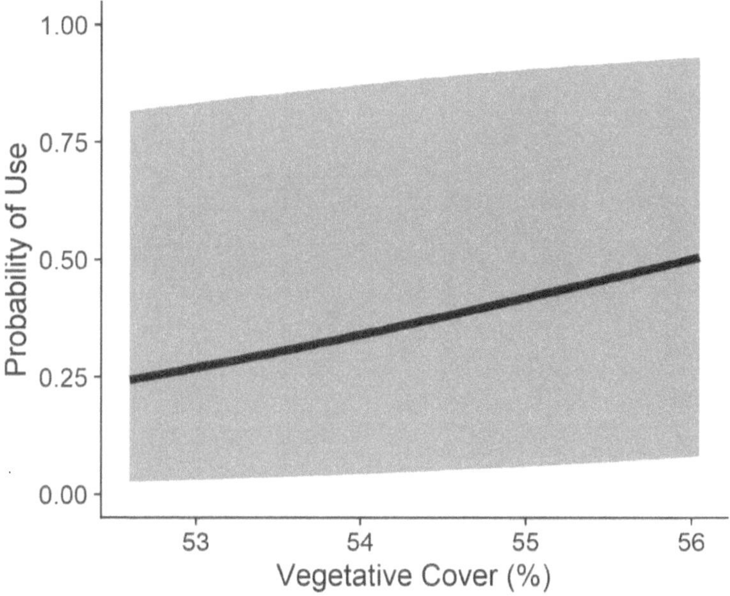

Figure 2.3. The effect of percent vegetative cover (< 1 m height) on the probability of site use by timber rattlesnakes in southern Ohio. Timber rattlesnakes were more likely to use sites with more vegetative ground cover. Shaded areas represent 95% credible intervals.

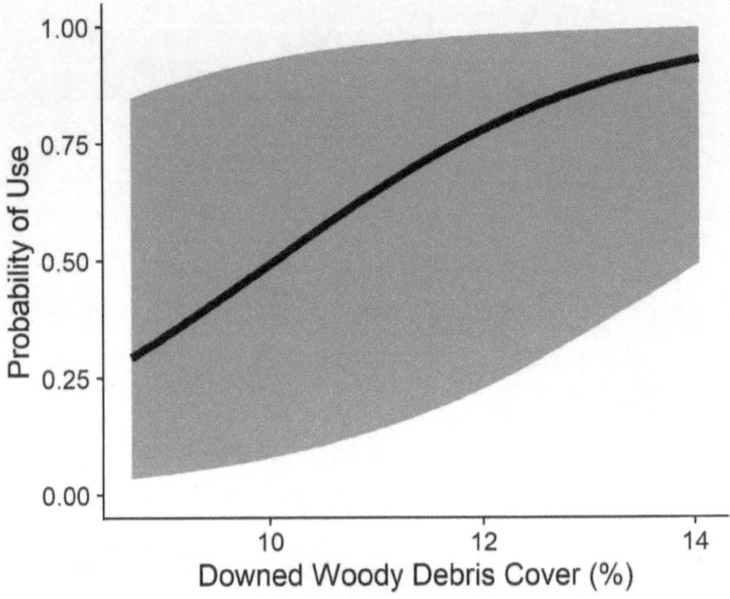

Figure 2.4. The effect of percent downed woody debris cover on the probability of site use by timber rattlesnakes in southern Ohio. Timber rattlesnakes were more likely to use sites with more downed woody debris. Shaded areas represent 95% credible intervals.

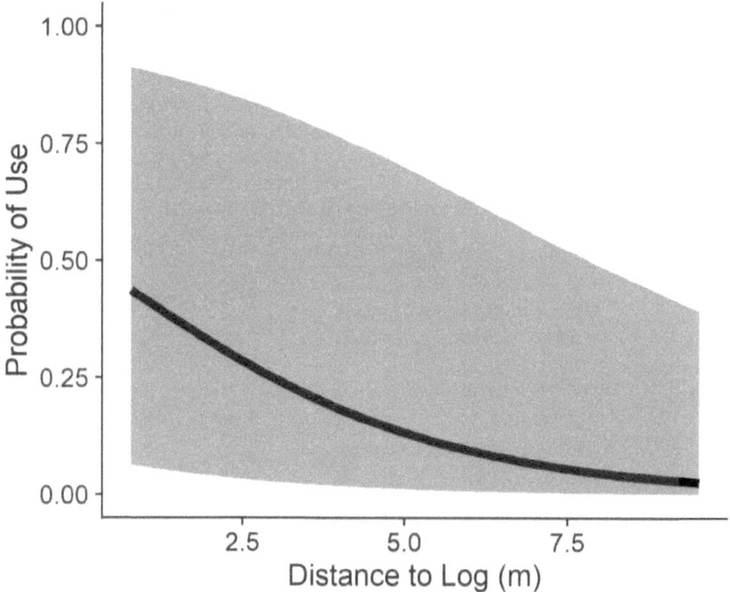

Figure 2.5. The effect of distance to nearest fallen log on the probability of site use by timber rattlesnakes in southern Ohio. Timber rattlesnakes were more likely to use sites nearer to fallen logs. Shaded areas represent 95% credible intervals.

Table 2.1. Estimated effects of six covariates on the probability of site use, in our forest structure and composition model, by timber rattlesnakes (*Crotalus horridus*) in southern Ohio. The lower and upper 95% highest density intervals (HDI-low and HDI-high respectively) are presented alongside the probability of direction (PD), and percent of the posterior distribution inside of the Region of Practical Equivalence (ROPE) using 89% of the posterior distribution. Bolded parameters represent covariates with < 15% of their posterior distribution inside ROPE. We included the Ecological Land Type Phases land cover class as a factor where the ridges (RID), SW upper slope (USW), NE upper slopes (UNE), SW lower slopes (LSW), and NE lower slopes (LNE) categories were compared against the bottomland (BTM) category individually.

Parameter	Estimate	HDI-high	HDI-low	PD	ROPE
Total DBH	0.15	0.38	-0.07	0.86	0.59
Total Basal Area	0.06	0.28	-0.16	0.67	0.85
Relative Density	**-0.47**	**-0.24**	**-0.71**	**1.00**	**0.00**
White Oak Importance	0.06	0.27	-0.13	0.69	0.87
Shannon's Div. Index	0.07	0.28	-0.14	0.71	0.84
RID-BTM	**1.15**	**1.70**	**0.57**	**1.00**	**0.00**
USW-BTM	**1.15**	**1.76**	**0.56**	**0.99**	**0.00**
UNE-BTM	**1.00**	**1.62**	**0.28**	**0.98**	**0.00**
LSW-BTM	**0.89**	**1.53**	**0.24**	**0.98**	**0.00**
LNE-BTM	0.41	1.25	-0.42	0.79	0.22

Table 2.2. Estimated effects of eight covariates on the probability of site use, in our micro-habitat model, by timber rattlesnakes (*Crotalus horridus*) in southern Ohio. The lower and upper 95% highest density intervals (HDI-low and HDI-high respectively) are presented alongside the probability of direction (PD), and percent of the posterior distribution inside of the Region of Practical Equivalence (ROPE) using 89% of the posterior distribution. Bolded parameters represent covariates with < 15% of their posterior distribution inside ROPE.

Parameter	Estimate	HDI-high	HDI-low	PD	ROPE
% Vegetative Cover	**0.33**	**0.49**	**0.18**	**1.00**	**0.00**
% Woody Debris Cover	**0.64**	**0.82**	**0.46**	**1.00**	**0.00**
% Rock Cover	-0.08	0.07	-0.24	0.83	0.90
% Tree Cover	0.19	0.32	0.05	0.99	0.49
Distance to Overstory Tree	-0.01	0.14	-0.16	0.56	1.00
Distance to Understory Tree	-0.03	0.14	-0.19	0.61	0.99
Distance to Log	**-0.38**	**-0.18**	**-0.58**	**1.00**	**0.01**
% Canopy Cover	0.04	0.20	-0.13	0.64	0.97

Chapter 3: Modeling ingress and egress of timber rattlesnakes (*Crotalus horridus*) to predict their exposure to prescribed fire

Introduction

Understanding phenology in wildlife populations is important, especially in ecosystems under pressure from climate change and anthropogenic disturbance. Phenology is the study of the natural activities and cycles of plants and animals and is concerned primarily with the timing and drivers of recurring events (Lieth 2013). Ecosystem and plant phenology has become increasingly studied as the global climate shifts and temperatures warm, causing changes in when plants advance through developmental phases during spring and fall (Badeck et al. 2004). Generally, in the northern hemisphere, bud burst, flowering, and green-up are advancing earlier in the spring although there is substantial regional variation (Peñuelas and Filella 2009, Piao et al. 2019). Climate change is also driving phenological shifts in bird migration (Cotton 2003, Gordo 2007) and mammal hibernation (Barnes et al. 2011, Pigeon et al. 2016, Delgado et al. 2018), but its impacts on smaller, more cryptic animals are still poorly understood. A growing body of literature on trends in amphibian breeding indicates many species in temperate regions are breeding earlier in the spring (Blaustein et al. 2001, Blaustein et al. 2002, Greene 2017). Although the effects of climate change on snakes and lizards are relatively understudied, there is mounting evidence that climate change is likely to impact these reptiles broadly, but variably (Gibbons et al. 2000, Winter et al. 2016).

To better understand climate-related risks, it is important to understand the timing and drivers of key phenological events in populations of threatened or declining reptiles. In temperate regions, snakes must brumate or undergo a hibernation-like state of inactivity during cold weather, to survive the winter months (Nordberg and Cobb 2017). Their spring emergence and dispersal away from hibernacula is referred to as egress, and the return to hibernacula in the fall is termed ingress (Brown 1992, Nordberg and Cobb 2017, Burger 2019). Timber rattlesnakes (*Crotalus horridus*) are a broadly distributed but rapidly declining species of snake found throughout eastern North America and, in many places, they are highly dependent on long-term, stabile overwintering sites and large communal dens (Galligan and Dunson 1979, Martin 2002). Ingress and egress represent critical periods for these snakes and must be timed precisely to avoid freezing temperatures (Shine and Mason 2004) and minimize exposure to predators at open-canopy sites (Herr et al. 2020), especially when snakes are sluggish. Although the thermal, and behavioral dynamics surrounding den use are well-studied (Brown 1982, Brown 1992, Nordberg 2013, Nordberg and Cobb 2017), predictive models of egress and ingress have not been published. Such a model would not only be useful in predicting future effects of climate change, but also in guiding forest management, especially pertaining to the use of prescribed fire.

Prescribed fire is a common management tool across many of the forests that timber rattlesnakes inhabit and is usually applied during the spring or fall, generally around the time snakes emerge from or return to dens (Knapp et al. 2009, Beaupre and Douglas 2012). These prescribed burns have the potential to cause extensive snake mortality if poorly timed (Beaupre and Douglas 2012), but fire is an important source of forest disturbance that can help maintain open habitats frequently used by gestating timber rattlesnakes (Hoffman et al. 2020) and has generally positive impacts on thermoregulation in reptiles (Russell et al. 1999). Fire is also an important tool used to combat mesophication, or the gradual increase in moisture and shade in the absence of forest disturbance due to increasing dominance of dense-canopied, shade tolerant trees, like beech or maple, and subsequent loss of shade-intolerant species like many oaks

(Abrams 1992, Nowacki and Abrams 2008, Alexander et al. 2021). Oaks are among the most

important and, historically, dominant trees in forests throughout the range of timber rattlesnakes

(Hanberry and Nowacki 2016), and oak acorns are an important food source for wildlife. In fact,

oak masting cycles can drive population fluctuations of the small mammals timber rattlesnake

feed on and may consequently impact rattlesnake populations (McShea and Schwede 1993, Elias

et al. 2004, Olson et al. 2015).

Prescribed fire is periodically applied as a management tool in southern Ohio forests

where the state's few remaining timber rattlesnake populations persist despite dramatic declines

elsewhere in the state (ODW 2011, Wynn 2012). Concerns over the use of prescribed fire on

these properties have arisen in the past, but the cryptic nature of snakes and the unpredictability

of opportune fire conditions makes a robust experimental study of fire's direct impact on snakes

impractical. We used multi-year radio-telemetry observations and game cameras placed on

known den sites alongside locally measured meteorologic data to generate a predictive model of

spring egress and fall ingress for timber rattlesnakes in southern Ohio. We hypothesized that

egress and ingress would be strongly driven by fluctuations in temperature, and that growing

degree day (GDD), a phenological measure of seasonally increasing cumulative mean daily

temperature, would be a good predictor of rattlesnake emergence, as it is for phenological events

with plants and insects (Miller et al. 2001, Herms 2004). Currently state agencies in Ohio may

not conduct prescribed burns on state properties after April 15th or before October 15th although

more stringent limitations have been proposed for properties where timber rattlesnakes occur.

Our goal was to model the risk of exposure to rattlesnakes in a way that allowed for decisions

regarding prescribed fire on these properties to be more nuanced and precise.

Methods

Study Site

We conducted our research at Vinton Furnace State Experimental Forest (VFSF), a

4,892-ha property located in Vinton County, Ohio, USA. VFSF is part of the Southern

Unglaciated Allegheny Plateau Ecological section, a primarily forested region typified by

heavily dissected topography (Cleland et al. 2007). Forests here are largely second-growth stands

recovering from exploitation during the 1800s (Stout, 1933), now actively managed to promote

habitat heterogeneity and oak regrowth. On-site research and management includes even- and

uneven-aged silvicultural treatments and prescribed fire (Ducey 1982).

Telemetry

We surgically implanted captured timber rattlesnakes with radio transmitters (Holohil SI-

2T) following procedures outlined by Reinert and Cundall (1982), using machine-administered

isoflurane to anesthetize animals via inhalation. Transmitter weight never exceeded 5% of the

snake's body weight, and in most instances was < 2%. We released snakes at their original

capture location within 24 hours of surgery and relocated individuals 2–3 times per week from

April through October 2016–2020. Upon visually locating a snake, we recorded the location with

a Global Positioning System (Garmin GPSmap 64s) to an estimated < 5m spatial accuracy. We

typically began checking dens during the spring on April 1st and stopped tracking telemetered

snakes during the fall after locating them underground at their den sites.

For each snake observation during egress (prior to June 1st) and ingress (after October

31st), we used detailed field notes to categorize each snake as either exposed (1) or sheltered (0).

We categorized snakes as sheltered if they were underground, inside a log, in a rock crevice, under a rock, or coiled within one meter of their den or other known shelter spot. In all other cases, we categorized snakes as exposed. Snakes were almost always near downed woody debris, so we only classified them as sheltered if they were coiled immediately outside of a known shelter they were observed using at other times. Although we are confident sheltered snakes would be at much lower risk from fire, some snakes classified as exposed would potentially have a means to reduce their exposure to or escape approaching fire; our model is therefore conservative and may overestimate risk.

Remote Cameras

We selectively deployed two models of trail cameras at dens during both the spring and the fall, usually attached to trees, stumps, or rocks to avoid driving stakes into the ground and potentially into dens. We used Day 6 Outdoors brand game cameras (TLC-200-C PlotWatcher) to capture time-lapse imagery over long periods of time, especially at relatively remote or inaccessible sites. These cameras only needed to be checked monthly but were only able to capture daytime imagery. Conversely, we used Moultrie Feeders brand game cameras (M-888 Mini Game Cams) cameras to capture time-lapse imagery during both day and night at most den sites. These cameras had to be checked multiple times per week and were therefore not practical for deployment at more remote sites. Placing cameras on dens with dense vegetation or steep topography, few trees, and thin soil was impractical, so some dens were not monitored via camera.

We used 30 second time-lapse intervals, as this was sufficient to capture the movement of snakes in and out of their dens while preserving battery life to allow for longer deployment. We tested the functionality of the motion-trigger setting on these cameras but found that, when paired with cameras set on time-lapse, cameras often did not trigger when snakes entered and exited dens due to their slow movements.

We typically deployed cameras during the first week of April and again during the first week of October, taking them back down when all telemetered snakes had exited or entered their dens, respectively. Given the imperfect record provided by PlotWatchers (no nighttime imagery), occasional camera malfunctions, and the numerous entry holes present at most dens, these cameras provide us with an imperfect record of den activity during the spring and fall. We therefore used these data to reduce uncertainty and increase our precision in defining when telemetered snakes entered or left their dens and did not attempt to quantify ingress and egress patterns of non-telemetered snakes captured on camera.

Meteorological Data

We collected weather data from an established Western Regional Climate Center Remote Automatic Weather Station (RAWS) in the middle of our study site. For our purposes, we used daily summaries of mean air temperature, mean relative percent humidity, mean wind speed, total daily precipitation, and daily high and low air temperatures. We used daily mean air temperature values to calculate GDD using a the widely used base temperature of 5° C (Raulier and Bernier 2000, Herms 2004).

Data Analysis

We modeled potential exposure to prescribed fire for timber rattlesnakes at our study site by using a generalized linear mixed effects Bernoulli model in a Bayesian framework using the brms package in R (Bürkner 2016; R Core Team 2020). We considered exposure (sheltered vs. exposed) as the response variable with GDD, mean daily relative humidity, mean daily solar radiation, mean daily wind speed, and mean daily temperature as predictor variables, and allowed for random variation by year and individual snake. We excluded both high and low daily air temperatures as they were highly correlated with mean daily air temperature, but less predictive of ingress and egress. We ran models with four MCMC chains for 15,000 iterations with a warmup of 3,000 and no thinning. We visually inspected chains to confirm adequate mixing, and we confirmed convergence using the Gelman-Rubin statistic (all Rhat values=1). We used the Region of Practical Equivalence (ROPE) defined as ±0.1*SD of the response variable (Kruschke 2018) to guide our interpretation of the most influential covariates in each model and considered variables for which <15% of the posterior distribution was inside of ROPE to be particularly influential. We then used this model and weather data from twelve dates from 2004–2016 where prescribed fire was applied on properties with known rattlesnake populations to predict the risk of fire exposure for timber rattlesnakes.

Results

We tracked 43 snakes to 28 hibernacula sites across the landscape (mean of 1.6 telemetered snakes per hibernacula). Rattlesnakes at our study site rarely emerged from their hibernacula prior to April 1st (day of year 90) and rarely returned to dens prior to September 1st (day of year 245), although there was considerable variation in timing over the subsequent two months for egress (Figure 3.1) and ingress (Figure 3.2) respectively. From 2017–2020, a mean proportion of 34.3% ± 16.5% SD of telemetered snakes had emerged from hibernacula by the close of the state's spring burn window (April 15th). From 2016–2020, a mean proportion of 52.1% ± 23.5% SD of telemetered snakes remained aboveground at the start of the state's fall burn window (October 15th).

We were able to track 33 of our 43 telemetered snakes to hibernacula over at least two years, allowing us to assess site fidelity. Of the 33 snakes monitored during two to four winters, five snakes switched dens a single time (mean site fidelity = 90.4 % ± 26 % SD). All five of these snakes were males and we were able to confirm that four successfully overwintered, but one died just weeks after emerging due to complications related to snake fungal disease. The fifth snake's transmitter presumably died shortly after ingress so we could not confirm his fate. In most cases, the second den was over 1 km from the first den used by a snake (den relocation distance mean = 848.4 m, range = 91–1,410 m).

Meteorologic covariates for both our ingress and egress models had similar predictive power in discriminating how likely a given snake was to be aboveground and exposed (Table 3.1, 3.2). GDD was the single best predictor of snake exposure during both ingress and egress; snakes were more likely to be aboveground and exposed with increasing GDD during the spring and less likely to be aboveground and exposed with increasing GDD in the fall (Figure 3.3, 3.4). Snakes were also more likely to be exposed during both spring and fall with increasing mean daily temperature (Figure 3.3, 3.4). Using the model fit to our five years of telemetry data, we predicted a mean risk of fire exposure of 33.4 % ± 13.9 % SD (range =5 %–52 %) for timber rattlesnakes located within a burn unit during the 12 most recent spring prescribed fires occurring on properties with known rattlesnake populations in southern Ohio (there was only a single fall prescribed fire during this time).

Discussion

Our telemetered snakes used twenty-five different den sites, some of which were apparently only used by a single snake. Most dens were nondescript, small holes with no associated aboveground rock, although a few were in more typical rock outcroppings on steep

slopes. Given the diffuse and diverse presentation of hibernacula at our study site, and that

snakes occasionally switched dens year to year over great distances, den sites may not be a

limiting factor on the landscape for timber rattlesnakes in Ohio. We also saw very little winter

mortality and only ever in association with emaciated or sickly snakes. Rattlesnakes at high

elevations and latitudes typically den communally at discrete dens with narrowly defined

topographic and thermal characteristics (Martin 2002) whereas populations toward the southern

extreme of the species' distribution often brumate singly and in scattered refugia (Waldron et al.

2006). Our observations indicate that timber rattlesnakes in Ohio exhibit an intermediary trend

by denning in small groups or singly at numerous dens across the landscape, similar to those in

the piedmont of North Carolina (Sealy 2002).

Spring egress typically began around April 1st, and progressed steadily, but variably

depending on temperature fluctuations with drops in temperature delaying emergence and

warming trends accelerating egress. The pattern was similar for ingress beginning in late

September, but slightly more variable in timing and with a much more deliberate and

synchronous movement during mid-October, around the time of the first freezing night. The last

day during spring and first day during fall of each year in which temperatures fell below freezing

were good markers for the beginning of egress and end of ingress, respectively. Although most

snakes remained underground until the end of the spring burn window (April 15th), about half of

our telemetered snakes, on average, were still aboveground at the start of the fall burning

window (October 15th). However, this number rapidly decreased over the following two weeks

with almost all snakes underground by November 1st. As with den site characteristics and

distribution, timing of egress and ingress in Ohio more closely matches that of central North

Carolina, where egress peaks in mid-April and ingress peaks during mid-October and is intermediary to trends observed in populations in the Appalachian Mountains and Coastal Plain (Martin 2002, Sealy 2002, Waldron et al. 2006).

We were further able to refine these observations by developing a predictive model of snake "exposure risk" wherein snakes underground or in a large log or rock outcrop were considered sheltered (and presumed to have a high probability of surviving a prescribed burn), and snakes aboveground and away from shelter were considered "exposed". Snake risk was best predicted by GDD and, to a lesser extent, by temperature. Our model predicted that a snake's probability of exposure rose rapidly starting at around GDD 250 and approached 100% around GDD 750, at which point most snakes were surface active, with fluctuations in temperature mediating this trend. Risk of exposure during the fall remained relatively high until GDD 2,800, at which point it dropped rapidly.

We used this model to retrospectively predict the risk of fire exposure for snakes during spring egress for the 12 most recent years where prescribed burns occurred on properties with known timber rattlesnake populations in Ohio and predicted a low to moderate risk of fire exposure for individual snakes within burn units on these days. However, given that snakes brumate diffusely across the property and that prescribed fires are conducted infrequently in relatively small areas, we speculate that this low to moderate risk of exposure for individual snakes likely translates to very infrequent rattlesnake mortalities and none have yet been observed.

This is not surprising, given that all burns occurred before April 16th, prior to most snakes leaving their dens and when those that have left are often near or in burrows, logs, or other shelters. Taken collectively, our observations and model indicate that current prescribed fire practices likely pose a low risk to timber rattlesnake populations in southern Ohio. An additional observation worth noting is that we monitored a den with multiple large adult rattlesnakes situated within one of the most frequently burned units on the property (4 burns conducted between 2001 and 2016) which was also located within a several hectare area that was recently exposed to a high severity fire with extensive overstory mortality. Snakes only infrequently switch dens, and most of the adults at this den had likely survived multiple fires burning over their hibernacula, indicating some level of resilience to fire, even when dens are burned directly over. Further, although we know little about behavioral fire responses in reptiles, there is evidence that snakes can respond to and mitigate their risk from fire (Russell et al. 1999, Smith et al. 2001). We believe our results can help guide forest management and inform prescribed fire protocol which has had to operate under very little guidance regarding timber rattlesnakes until now.

Our results also indicate some level of flexibility for rattlesnakes to adjust the timing of ingress and egress based on changes in temperature which may result in altered phenology in association with climate change, as has been show in Mediterranean vipers (Rugiero et al. 2013). Future studies should seek to better understand how flexible ingress and egress are across a broader time frame with more extreme fluctuations in temperature. Erratic or extreme weather patterns might disrupt natural cycles and lead to long-term decline as has been implied for massasaugas (*Sistrurus catenatus*), and climate change has been proposed as a contributing factor to the decline of timber rattlesnakes in New Hampshire (Clark et al. 2011, Pomara et al. 2014). This study introduces as many questions as it answers, but our model and multi-year observations will serve as an important framework for better understanding the role of temperature and day of year on the phenology of brumation in snakes.

Literature Cited

Abrams, M.D. 1992. Fire and the development of oak forests. BioScience 42.5:346–353.

Badeck, F.W., A. Bondeau, K. Böttcher, D. Doktor, W. Lucht, J. Schaber, and S. Sitch. 2004.

Responses of spring phenology to climate change. New phytologist 162.2:295–309.

Alexander, H.D., C. Siegert, J.S. Brewer, J. Kreye, M.A. Lashley, J.K. McDaniel, A.K. Paulson,

H.J. Renninger, and J.M. Varner. 2021. Mesophication of Oak Landscapes: Evidence,

Knowledge Gaps, and Future Research. BioScience.

Barnes, B., M. Sheriff, J. Kenagy, L. Buck, and T. Squirrel. 2011, December. The influence of

changing seasonality and snow cover on arctic ground squirrel phenology. In AGU Fall

Meeting Abstracts (Vol. 2011, B52–B02).

Beaupre, S.J. and L.E. Douglas. 2012. October. Responses of timber rattlesnakes to fire: lessons

from two prescribed burns. Proceedings of the 4th Fire in Eastern Oak Forests

Conference:192–204.

Blaustein, A.R., L.K. Belden, D.H. Olson, D.M. Green, T.L. Root, and J.M. Kiesecker. 2001.

Amphibian breeding and climate change. Conservation Biology 15.6:1804–1809.

Blaustein, A.R., L.K. Belden, and D.H. Olson. 2002. Amphibian phenology and climate change.

Conservation Biology 16.6:1454–1455.

Brown, W.S. 1982. Overwintering body temperatures of timber rattlesnakes (*Crotalus horridus*)

in northeastern New York. Journal of Herpetology 16:145–150.

Brown, W.S. 1992. Emergence, ingress, and seasonal captures at dens of northern timber

rattlesnakes, *Crotalus horridus*. *In* Biology of the Pitvipers (ed. E.D. Brodie) 251–258.

Burger, J. 2019. Vulnerability of Northern Pine Snakes (*Pituophis melanoleucus* Daudin, 1803)
during fall den ingress in New Jersey, USA. Amphibian & Reptile
Conservation, 13.2:102–114.

Clark, R.W., M.N. Marchand, B.J. Clifford, R. Stechert, and S. Stephens. 2011. Decline of an
isolated timber rattlesnake (*Crotalus horridus*) population: interactions between climate
change, disease, and loss of genetic diversity. Biological Conservation 144.2:886–891.

Cleland, D.T., J.A. Freeouf, J.E. Keys, G.J. Nowacki, C.A. Carpenter, W.H. McNab. 2007.
Ecological Subsections: Sections and Subsections for the Conterminous United States;
Gen. Tech. Report WO-76D [Map on CD-ROM], Sloan, A.M. cartographer, presentation
scale 1:3,500,000, colored; Department of Agriculture, Forest Service: Washington, DC,
USA.

Cotton, P.A. 2003. Avian migration phenology and global climate change. Proceedings of the
National Academy of Sciences 100.21:12219–12222.

Delgado, M.M., G. Tikhonov, E. Meyke, M. Babushkin, T. Bespalova, S. Bondarchuk, A.
Esengeldenova, I. Fedchenko, Y. Kalinkin, A. Knorre, and G. Kosenkov. 2018. The
seasonal sensitivity of brown bear denning phenology in response to climatic
variability. Frontiers in zoology 15.1:1–11.

Ducey, L.J.R., 1982. Hardwood research in southern Ohio. Journal of Forestry, 80.8:488–489.

Elias, S.P., J.W. Witham, and M.L. Hunter. 2004. *Peromyscus leucopus* abundance and acorn
mast: population fluctuation patterns over 20 years. Journal of Mammalogy 85.4:743–
747.

Galligan, J.H. and W.A. Dunson. 1979. Biology and status of timber rattlesnake (*Crotalus
horridus*) populations in Pennsylvania. Biological Conservation 15.1:13–58.

Gibbons, J.W., D.E. Scott, T.J. Ryan, K.A. Bulhmann, T.D. Tuberville, B.S. Metts, J.L. Greene, T. Mills, Y. Leiden, S. Poppy, and C.T. Winne. 2000. The global decline of reptiles, déjà vu amphibians. Bioscience 50.8:653–666.

Green, D.M. 2017. Amphibian breeding phenology trends under climate change: predicting the past to forecast the future. Global change biology 23.2:646–656.

Gordo, O. 2007. Why are bird migration dates shifting? A review of weather and climate effects on avian migratory phenology. Climate research 35.1–2:37–58.

Hanberry, B.B. and G.J. Nowacki. 2016. Oaks were the historical foundation genus of the east-central United States. Quaternary Science Reviews 145:94-103.

Herms, D.A. 2004. Using degree-days and plant phenology to predict pest activity. In IPM (integrated pest management) of Midwest landscapes 58:49–59.

Herr, M.W., J.D. Avery, T. Langkilde, and C.A. Howey. 2020. Trade-off between thermal quality and predation risk at Timber Rattlesnake gestation sites. Journal of Herpetology 54.2:196-205.

Hoffman, A.S., A.M. Tutterow, M.R. Gade, B.T. Adams, and W.E. Peterman. 2020. Variation in behavior drives multiscale responses to habitat conditions in timber rattlesnakes (Crotalus horridus). BioRxiv [**Preprint**]. December 4, 2020 [cited 2020 Dec 31]. Available from: https://doi.org/10.1101/2020.12.03.410290.

Knapp, E.E., B.L. Estes, and C.N. Skinner. 2009. Ecological effects of prescribed fire season: a literature review and synthesis for managers. USDA General Technical Report PSW-GTR-224.

Kruschke, J.K. 2018. Rejecting or accepting parameter values in Bayesian estimation. Advances in Methods and Practices in Psychological Science 1.2:270–280.

Lieth, H. ed. 2013. *Phenology and seasonality modeling* (Vol. 8). Springer Science & Business.

Martin, W.H. 2002. Life history constraints on the timber rattlesnake (*Crotalus horridus*) at its climatic limits. Biology of the Vipers 285–306.

McShea, W.J. and G. Schwede. 1993. Variable acorn crops: responses of white-tailed deer and other mast consumers. Journal of Mammalogy 74.4:999–1006.

Media. Reinert, H.K. and D. Cundall. 1982. An improved surgical implantation method for radio-tracking snakes. Copeia 3:702–705.

Miller, P., W. Lanier, and S. Brandt. 2001. Using growing degree days to predict plant stages. Ag/Extension Communications Coordinator, Communications Services, Montana State University-Bozeman, Bozeman, MO, 59717 406:994–2721.

Nordberg, E.J. 2013. Thermal ecology and behavioral activity in hibernating timber rattlesnakes (*Crotalus horridus*) Doctoral book, Middle Tennessee State University.

Nordberg, E.J. and V.A. Cobb. 2017. Body temperatures and winter activity in overwintering timber rattlesnakes (*Crotalus horridus*) in Tennessee, USA. Herpetological Conservation and Biology 12.3:606–615.

Nowacki, G.J. and M.D. Abrams. 2008. The demise of fire and "mesophication" of forests in the eastern United States. BioScience 58.2:123–138.

Ohio Division of Wildlife (ODW). 2011. Timber Rattlesnake (*Crotalus horridus*). Ohio Department of Natural Resources. *<wildlife.ohiodnr.gov/species-and-habitats/species-guide-index/reptiles/timber-rattlesnake>*

Olson, Z.H., B.J. MacGowan, M.T. Hamilton, A.F. Currylow, and R.N. Williams. 2015. Survival of Timber Rattlesnakes (*Crotalus horridus*): Investigating individual, environmental, and ecological effects. Herpetologica, 71.4:274–279.

Peñuelas, J. and I. Filella. 2009. Phenology feedbacks on climate change. Science 324.5929:887–888.

Piao, S., Q. Liu, A. Chen, I.A. Janssens, Y. Fu, J. Dai, L. Liu, X. Lian, M. Shen, and X. Zhu. 2019. Plant phenology and global climate change: Current progresses and challenges. Global change biology 25.6:1922–1940.

Pigeon, K.E., G. Stenhouse, and S.D. Côté. 2016. Drivers of hibernation: linking food and weather to denning behaviour of grizzly bears. Behavioral Ecology and Sociobiology 70.10:1745–1754.

Pomara, L.Y., O.E. LeDee, K.J. Martin, and B. Zuckerberg. 2014. Demographic consequences of climate change and land cover help explain a history of extirpations and range contraction in a declining snake species. Global Change Biology 20.7:2087–2099.

Raulier, F. and P.Y. Bernier. 2000. Predicting the date of leaf emergence for sugar maple across its native range. Canadian journal of forest research 30.9:1429–1435.

Rugiero, L., G. Milana, F. Petrozzi, M. Capula, and L. Luiselli. 2013. Climate-change-related shifts in annual phenology of a temperate snake during the last 20 years. Acta oecologica, 51:42–48.

Russell, K.R., D.H. Van Lear, and D.C. Guynn. 1999. Prescribed fire effects on herpetofauna: review and management implications. Wildlife Society Bulletin 27.2:374–384.

Sealy, J.B. 2002. Ecology and behavior of the timber rattlesnake (*Crotalus horridus*) in the upper Piedmont of North Carolina: Identified threats and conservation recommendations. *In* Biology of the Vipers (ed. G.W. Schuett) 561–578.

Shine, R. and R.T. Mason. 2004. Patterns of mortality in a cold-climate population of garter snakes (*Thamnophis sirtalis parietalis*). Biological Conservation 120.2:201–210.

Smith, L.J., A.T. Holycross, C.W. Painter, and M.E. Douglas. 2001. Montane rattlesnakes and

prescribed fire. The Southwestern Naturalist 46.1:54–61.

Stout, W. 1933. The charcoal iron industry of the Hanging Rock Iron District—its influence on

the early development of the Ohio Valley. Ohio State Archaeol. Hist. Q. 42, 72–104.

Waldron, J.L., J.D. Lanham, and S.H. Bennett. 2006. Using behaviorally-based seasons to

investigate canebrake rattlesnake (*Crotalus horridus*) movement patterns and habitat

selection. Herpetologica 62.4:389–398.

Winter, M., W. Fiedler, W.M. Hochachka, A. Koehncke, S. Meiri, and I. De la Riva. 2016.

Patterns and biases in climate change research on amphibians and reptiles: a systematic

review. Royal Society Open Science 3.9:160158.

Wynn, D. 2012. Timber rattlesnake populations in Ohio forests. Presentation—Timber

Rattlesnake Workshop, Vinton-Furnace State Forest.

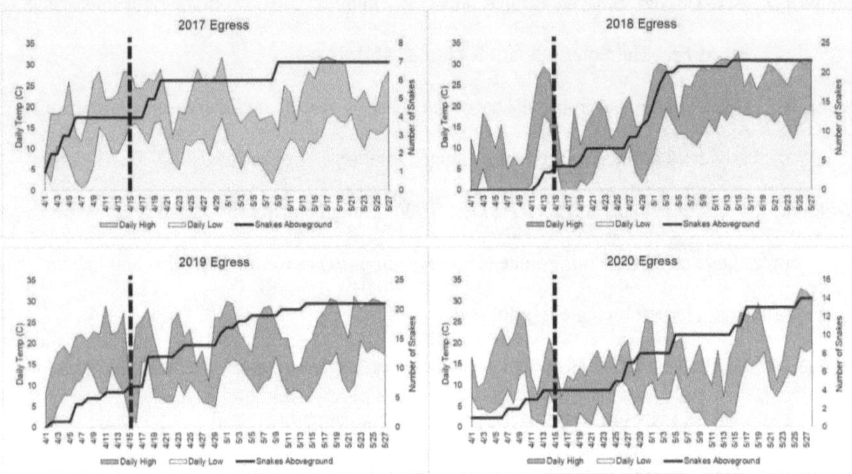

Figure 3.1. Cumulative emergence of radio-telemetered timber rattlesnakes (*Crotalus horridus*) from multiple hibernacula in southern Ohio during April and May across four years. The solid black line represents the number of snakes known to have emerged from hibernacula on a given day and the dashed lines represents the current April 15th cutoff for burning on state properties in Ohio. Although the number of telemetered snakes monitored varied by year (see secondary axis), the cumulative trend lines displayed are directly comparable between years as they effectively illustrate the proportion of monitored snakes currently above ground at any given time.

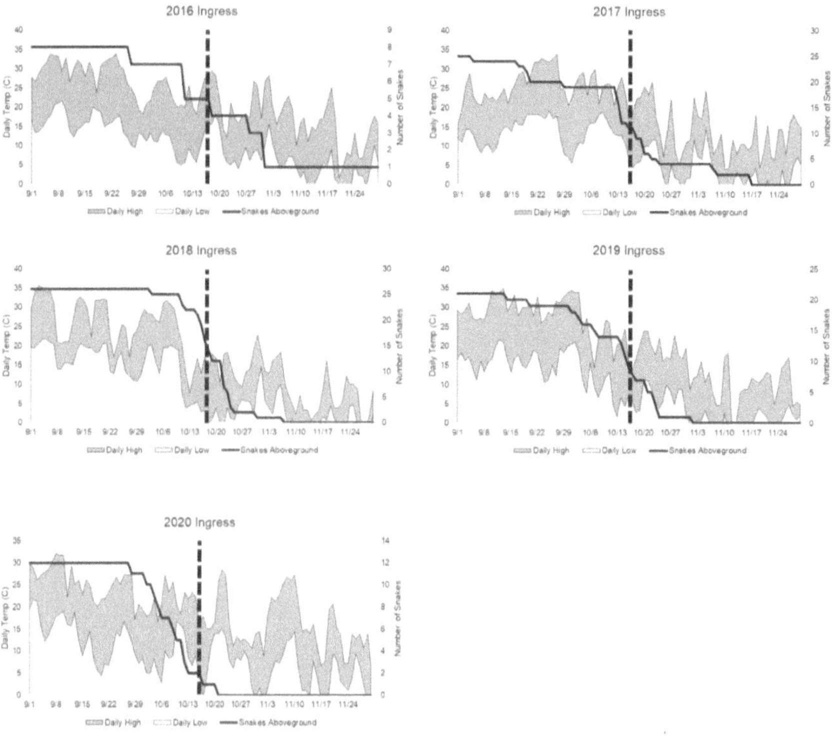

Figure 3.2. Cumulative ingress of radio-telemetered timber rattlesnakes (*Crotalus horridus*) to multiple hibernacula in southern Ohio during April and May across five years. The solid black line represents the number of snakes known to still be aboveground on a given day and the dashed lines represents the current October 15th cutoff for beginning fall burns on state properties in Ohio. Although the number of telemetered snakes monitored varied by year (see secondary axis), the cumulative trend lines displayed are directly comparable between years as they effectively illustrate the proportion of monitored snakes currently above ground at any given time.

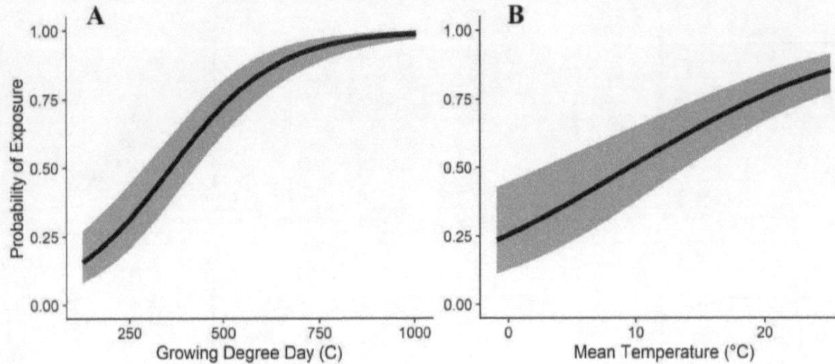

Figure 3.3. The effects of the two most influential covariates on the probability of fire exposure to timber rattlesnakes (*Crotalus horridus*) during spring egress in southern Ohio. The probability of exposure increases with increasing growing degree day (A) and mean daily temperature (B).

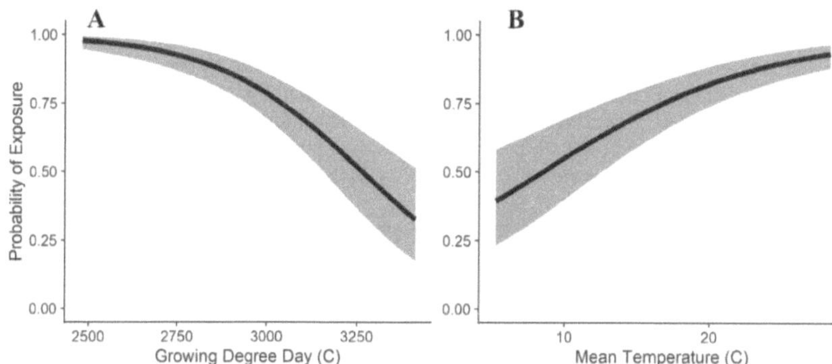

Figure 3.4. The effects of the two most influential covariates on the probability of fire exposure to timber rattlesnakes (*Crotalus horridus*) during fall ingress in southern Ohio. The probability of exposure decreases with increasing growing degree day (A) and increases with mean daily temperature (B).

Table 3.1. Estimated effects of environmental covariates on the probability of exposure to fire during spring egress by timber rattlesnakes (*Crotalus horridus*) in southern Ohio. We present the lower and upper 95% highest density intervals (HDI-low and HDI-high respectively), the probability of direction (PD), and percent of the posterior distribution inside of the Region of Practical Equivalence (ROPE) using 89% of the posterior distribution. Bolded parameters represent covariates with < 1% of their posterior distribution inside ROPE.

Parameter	Estimate	HDI-low	HDI-high	PD	ROPE
Intercept	0.77	0.39	1.16	1.00	0.00
Growing Degree Day	**1.57**	**1.24**	**1.90**	**1.00**	**0.00**
Relative Humidity	0.30	-0.03	0.53	0.93	0.31
Mean Temperature	**0.70**	**0.59**	**0.91**	**1.00**	**0.00**
Solar Radiation	0.19	-0.06	0.44	0. 89	0.47
Wind Speed	-0.25	-0.45	-0.05	0.98	0.27

Table 3.2. Estimated effects of environmental covariates on the probability of exposure to fire during fall ingress by timber rattlesnakes (*Crotalus horridus*) in southern Ohio. We present the lower and upper 95% highest density intervals (HDI-low and HDI-high respectively), the probability of direction (PD), and percent of the posterior distribution inside of the Region of Practical Equivalence (ROPE) using 89% of the posterior distribution. Bolded parameters represent covariates with < 1% of their posterior distribution inside ROPE.

Parameter	Estimate	HDI-low	HDI-high	PD	ROPE
Intercept	1.32	0.94	1.73	1.00	0.00
Growing Degree Day	**- 1.13**	**- 1.44**	**-0.83**	**1.00**	**0.00**
Relative Humidity	0.24	- 0.09	0.56	0.88	0.38
Mean Temperature	**0.82**	**0.59**	**1.04**	**1.00**	**0.00**
Solar Radiation	- 0.11	- 0.46	0.25	0.69	0.60
Wind Speed	- 0.08	- 0.25	0.10	0.75	0.87

Chapter 4: Assessing social tolerance for rattlesnakes in Ohio with a psychological model and

multiple measures of tolerance

Introduction

Research increasingly reveals a link between extinction risk for wild animals and human

activities. Human threats are particularly relevant for species know to threaten humans, or the

things we value (e.g., livestock). These circumstances have led to rapid growth in interest of applying psychological models to better understand tolerance for wildlife—particularly large, predatory mammals (Zajac et al. 2012, Bruskotter and Wilson 2014, Lischka et al. 2019). The term "tolerance" has been applied to a broad range of phenomena (Treves and Bruskotter 2014) including attitudes toward species, intentions to engage in behaviors related to those species, as well as attitudes toward species management (Bruskotter and Fulton 2012, Slagle and Bruskotter 2019). Most commonly, tolerance is employed as a synonym for "acceptance" as measured by wildlife acceptance capacity (Decker and Purdy 1988, Kansky et al. 2016) or wildlife stakeholder acceptance capacity (Riley and Decker 2000) and is essentially a measure of an individual's willingness to accept risks or costs associated with the presence of a given wild animal (Bruskotter et al. 2015, Kansky et al. 2016, Lischka et al. 2019).

Evidence indicates that an individual's perception of the potential risks and benefits associated with the presence of potentially dangerous wildlife can strongly influence their tolerance for wild animals (Zajac et al. 2012, Bruskotter and Wilson 2014, Lischka et al. 2019) and their support for its conservation (Slagle et al. 2013). Tolerance for wildlife can also be influenced by one's wildlife value orientations (Zinn et al. 2000, Manfredo et al. 2009)—that is, their perceptions about the ideal relationships between humans and wildlife, as well as previous interactions with wild animals (Riley et al. 2002). Research suggests risk and benefit perceptions are the best predictors of acceptance for large mammalian predators (Lischka et al. 2019; Inskip et al. 2016, Bruskotter and Wilson 2014, Kansky et al 2016). Yet, while these models have established a critical framework for understanding human dimensions in conserving persecuted and declining large mammals, they have rarely been applied to smaller, less charismatic taxa, like snakes.

Fear of snakes is the most prevalent phobia worldwide (Agras et al. 1969; Ceríaco 2012) and this fear may be evolutionary in origin (DeLoache and LoBue 2009; Öhman and Mineka 2003). The ubiquity and strength of fear towards snakes can hamper conservation efforts aimed at recovering this group of animals in rapid, global decline (Gibbons et al. 2000, Reading et al. 2010). Seigel and Mullin (2009), emphasize that snakes are broadly threatened and that fear or intolerance toward snakes is a significant barrier to their conservation. While there is a growing database and consensus on the global decline of snakes, there is little information on the role widespread fear and persecution may play in this decline.

Studies from Portugal (Ceríaco 2012), Nepal (Pandey et al. 2016), Slovakia and Turkey (Prokop et al. 2010), and the United States (Agras et al. 1969) all found negative public attitudes towards snakes and snake conservation were widespread. Seigel and Mullin (2009) urged snake ecologists to become advocates for snake conservation but offered no science-based guidance on assessing or addressing barriers to acceptance. However, the relatively robust literature on tolerance with large mammals could be used to inform our assessment of the psychological mechanisms underlying negative perceptions of snakes.

Timber rattlesnakes (*Crotalus horridus*) in Ohio make for an appealing case study to better understand tolerance for snakes. Over the past century, timber rattlesnakes declined by at least 66% in Ohio (ODW 2011), but some evidence points to more than a 90% decline (Wynn, 2012). Though data on the scale of historic extermination efforts are scarce, direct persecution (i.e., human intolerance) was clearly a factor in this decline (Levin 2016). Given that these snakes were once legally harvested for bounties but are now protected in Ohio and elsewhere, the state of their conservation has clearly improved. However, the perception of snakes as hazards continues to be a real barrier to recovering species which have disappeared from most of their former range (Norris 2017), as it has with large mammalian carnivores (Kellert et al. 1996). In fact, The Ohio Division of Wildlife's current conservation efforts are directed at protecting "existing populations" of snakes—effectively excluding the potential for recovering extirpated rattlesnake populations within their former range (ODW 2011). Although the perceived risk from snakes far outweighs the actual danger venomous snakes present (Forrester et al. 2018), accurately conveying this level of risk will likely not be enough to promote broad social tolerance for snakes (Slagle et al. 2013).

To improve outreach efforts for threatened and endangered snakes, we must first understand the basis for resistance to their conservation. Following psychological models of 'hazard acceptance' used to predict tolerance for large predators (Zajac et al. 2012, Bruskotter and Wilson 2014, Lischka et al. 2019), we surveyed public perception of timber rattlesnakes in Ohio to better understand the mechanisms underlying tolerance for rattlesnakes and support for conservation efforts that could help ensure the long-term conservation of this species. Following these models, we hypothesized that an individual's acceptance of rattlesnakes would be affected by their perception of the potential risks and benefits of having rattlesnakes on the landscape which would be influenced by higher order attitudes and wildlife value orientations (Figure 4.1).

Beyond their acceptance of rattlesnakes, some research suggests that hazard models can predict people's intentions to engage in supportive (e.g., stewardship/conservation) or oppositional actions. Thus, we also assessed intentions to support rattlesnake conservation through the same conceptual framework. Though these three concepts (desired rattlesnake population, encounter-response scenarios, and support for conservation) are distinct, they are closely related (Bruskotter and Fulton 2012, Bruskotter et al. 2015) and we predicted that an individual's perceptions of risks and benefits associated with rattlesnakes would be similarly important in understanding all three variables.

Methods

Data Collection

We used a private sampling firm (Qualtrics XM, Inc., Seattle, WA) to gain access to a panel of Ohio adults and distributed an online survey to sampled individuals via the Qualtrics platform. Our primary aim was obtaining results reflective of a general cross section of Ohio's adult residents, rather than smaller or more specifically motivated groups (e.g., hunters, anglers) that are more likely to respond to traditional mail surveys (e.g., Thompson 1991). Qualtrics provides panelists recruited to participate with a cash-equivalent-incentive for completing the survey in order to improve response rates and ensure respondents completed the survey thoroughly, regardless of their interest level. Our survey began on 11/11/2020 and ran until 11/30/2020, when we collected our target 400 completed responses (total of 447 surveys completed).

Variables

<u>Wildlife value orientations</u>. We measured wildlife value orientations by asking study participants to indicate their level of agreement, on a 5-point bi-polar response scale (i.e., 2=agree, 1=somewhat agree, 0=neither agree nor disagree, -1=somewhat disagree, -2=disagree), with a subset of response items used by Manfredo et al. (2009) to measure mutualism and domination-based wildlife values. We selected four items from each category to ultimately measure mutualism and domination, respectively, based on factor loadings in our confirmatory analysis (Table 4.1).

Attitudes toward rattlesnakes. We measured general attitudes toward rattlesnakes by asking study participants to indicate their level of agreement, on a 5-point bipolar response scale (2=strongly agree, 1=somewhat agree, 0=unsure, -2=somewhat disagree, -1=strongly disagree), to belief statements about rattlesnakes, half of which have previously been used to measure *coexistence* and *concerns* pertaining to tolerance of rattlesnakes (Christoffel 2007, Keener-Eck 2017). We also measured study participants' attitudes toward various animals (rattlesnake, mosquito, cat, dog, etc.), on a 7-point bipolar response (3=strongly like, 2= like, 1=somewhat like, 0= neither like nor dislike, -1=somewhat dislike, -2=dislike, -3=strongly dislike), by asking how much they like or dislike each animal (see Kellert 1985). We used five items from these response items to measure attitudes toward rattlesnakes (Table 4.1).

Perception of Benefits & Risks. We used two response items related to potential benefits associated with rattlesnakes relating to ecosystem services to measure perceived benefits from rattlesnakes (Table 4.1). We selected four response items related to potential risks associated with rattlesnakes to measure perceived risks from rattlesnakes (Table 4.1).

Desired rattlesnake population. We measured respondents' desired rattlesnake population by asking study participants to indicate what they would consider an ideal population of rattlesnakes in Ohio on a four-point Likert scale (3=healthy and abundant population, frequent sightings, 2=small and isolated population, occasional sightings, 1=population risks extinction, sightings are rare, 0=no rattlesnakes). We also asked study participants to indicate the kind of population trajectory they would most prefer to see in Ohio rattlesnakes on a four-point scale that is slightly modified from the traditional "wildlife acceptance capacity measure" (1=increase, 0=stay about the same, -1=decrease, -2=go away). We used these two items for our first measure of tolerance: desired population (Table 4.1).

Encounter-response scenarios. We asked participants to indicate how they would respond to a series of hypothetical scenarios involving encounters with snakes (Christoffel 2007, Keener-Eck 2017). Response options included: do nothing (coded as 0), call local snake volunteer (coded as 0), call Ohio DNR (coded as 0), attempt to move the snake (coded as -1); call animal control or police (coded as -2), or kill the snake (coded as -3). We used all three items from this section that directly reference a scenario involving a venomous snake for our second measure of tolerance: encounter-response scenarios toward rattlesnakes (Table 4.1).

Support for rattlesnake conservation. We measured stewardship for rattlesnakes by asking study participants to indicate their level of support, on a 5-point bipolar response scale (2=strongly support, 1=somewhat support, 0=unsure, -1=somewhat against, -2=strongly against), to items used by Keener-Eck (2017) and two response items we generated dealing specifically with reintroduction and translocation efforts. We selected four items from this section that most directly deal with legal protections and translocation efforts of rattlesnakes as our third measure of tolerance: support for rattlesnake conservation (Table 4.1).

We conducted confirmatory factor analyses to describe scale dimensionality for all latent variables (Table 4.1). After removing two items with low factor loadings contributing to poor model fit, all multi-item measures were unidimensional with good model fit (CFI > 0.95) and each item had acceptable factor loadings ($\beta > 0.50$).

Data Analysis

We built and analyzed a structural equation model (SEM) using maximum likelihood estimation with the lavaan package in R (R Core Team 2020). We built one model and used it to predict three separate measures of tolerance (i.e., desired rattlesnake population, encounter-response scenarios, and support for conservation) for timber rattlesnakes. We assessed model fit with the Comparative Fit Index (CFI) and a parsimony-adjusted measure, the Root Mean Squared Error of Approximation (RMSEA). We considered CFI values over 0.90 and RMSEA values less than 0.08 to be indicative of an acceptable fit (Hu and Bentler 1999).

Results

Response

We collected a total of 447 completed surveys. All respondents in our study were residents of Ohio with a mean age of 47 and equal representation (50%) of respondents identifying as male or female. Roughly two-thirds (74%) of respondents had completed some post-secondary education and under half (41%) held a bachelor's degree or higher. All respondents reported engaging in at least one outdoor activity with most (80%) participating in consumptive outdoor activities (hunting, fishing, mushroom hunting, etc.) and nearly all (90%) participating in non-consumptive outdoor activities (hiking, kayaking, birdwatching, etc.). A very small number of respondents (6%) reported purposefully searching for reptiles and amphibians, or "herping", while over a quarter (27%) reported doing some birdwatching. Over a quarter of respondents (39%) reported having visited one of six public properties known to have harbored timber rattlesnake populations in recent years with 33 of these respondents (7% of total sample) reporting encounters with wild timber rattlesnakes in Ohio.

Structural Equation Models

We fit three models explaining each of the three measures of tolerance for rattlesnakes. Though all three models showed good model fit (CFI > 0.9, RMSEA < 0.08), risks and benefits were only significant predictors for one measure of tolerance: desired population (CFI = 0.94, RMSEA = 0.051; Figure 4.2). As predicted, an individual's perception of the risks posed by rattlesnakes had a negative effect on tolerance (β = -0.46) and their perception of associated

benefits had a positive effect on tolerance (β = 0.52). An individual's attitude toward rattlesnakes

had the strongest effect on both perceived risks (β = -0.63) and benefits (β = 0.63). Though

domination wildlife value orientations had a notable positive effect on perceived risks (β = 0.40),

mutualism-based value orientations had a weak positive effect on both perceived benefits (β =

0.11) and risks (β = 0.25).

Discussion

We modeled tolerance for timber rattlesnakes among Ohio residents and found that an

individual's perception of the risks and benefits associated with rattlesnake presence on the

landscape were strong predictors of their tolerance for rattlesnakes. To our knowledge, this is the

first test of the risks-benefits psychological model of tolerance (Lischka et al.2019), used

commonly with large predatory mammals (Zajac et al. 2012, Bruskotter and Wilson 2014), in

snakes. We also used wildlife value orientations (Zinn et al. 2000, Manfredo et al. 2009),

alongside attitudes toward rattlesnakes, to predict perceptions of risks and benefits. Though

mutualism was only modestly predictive of perceived benefits associated with rattlesnakes,

domination had a stronger effect on perceived risks, indicating that wildlife value orientations are

useful in understanding tolerance of rattlesnakes. We also characterized attitude using response

items that gauged respondents' level of positivity (e.g., "to what extent do you like or dislike this

animal?") and interest (e.g., "I would enjoy seeing a rattlesnake in the wild") toward rattlesnakes

and found that it influenced perceptions of risks and benefits even more than a person's wildlife

value orientations. Similarly, Keener-Eck et al. (2020) found that attitudes toward rattlesnakes specifically were better predictors of tolerant behavioral intentions toward rattlesnakes than wildlife value orientations.

However, our model was only significantly predictive of tolerance when measured as the desired population of rattlesnakes in Ohio. Perceptions of risks and benefits did not significantly impact either encounter-response scenarios toward rattlesnakes, or support for rattlesnake conservation efforts. In practice, 'tolerance' has been assessed using a variety of measures—from an individual's attitudes toward a particular species to over behaviors; however, these are rarely assessed in the same study (Bruskotter et al. 2015). Our findings indicate that, at least for some species, common indicators of tolerance (e.g., wildlife stakeholder acceptance capacity, encounter-responses, and support for conservation measures) are not closely related or affected by the same variables.

We believe this may be related to the disproportionate fear associated with snakes (Agras et al. 1969; Ceríaco 2012, Forrester et al. 2018), which may be relatively unrelated to the perception of actual risk posed by an animal. Venomous snakebites are a rare phenomenon in the United States (3,000–9,000 bites treated per year; O'Neil et al. 2007) and deaths caused by venomous snakes are even rarer (~ 5 deaths reported per year, Langley 2005). Venomous snakes in Ohio are found predominantly in the least populated, most rural parts of the state and rattlesnakes were reportedly encountered by less than 10% of our respondents. Yet about half of respondents expressed that rattlesnakes posed an unacceptable threat to either pets (48%) or children (53%), almost half (44%) expressed reluctance to visit a park where rattlesnakes occurred, and some (16%) expressed reluctance to visit a nature center housing a captive rattlesnake. A variety of studies indicate that risks and benefits can be relatively strongly associated with emotional responses (particularly affect). Here we used a measure of attitudes, as opposed to emotional measures, to capture one's general disposition toward snakes. As in prior studies, this measure was strongly related with perceptions of risk and benefit.

Humans assess and perceive risk through analytical (logic and rational thought) and experiential (emotions and associations) means (Slovic et al. 2004), but the latter often overshadows the former via a mechanism known as the affect heuristic (Slovic et al. 2007). This emotional reaction toward specific types of wildlife can be more important in determining tolerance for wildlife than the realized costs of living alongside an animal (Slovic 1982, Jacobs et al. 2012). Given that phobias, such as ophidiophobia, are emotionally based (Agras et al. 1969), acceptance or tolerance of snakes may be helped very little by fact-based messages (i.e., conveying how low an individual's risk of snakebite is).

The conservation of timber rattlesnakes in Ohio depends on both sound land management and sufficient public support to carry out conservation objectives. Our results indicate that many Ohio residents are in support of reintroducing rattlesnakes where populations have been extirpated (38% support, 28% against) and introducing more snakes into existing populations that are declining (33% support, 35% against) but about half of respondents would not enjoy encountering (45%) or living near rattlesnakes (51%). Given the disconnect between our three measures of tolerance, acceptance of rattlesnakes in the state likely does not equate to support for the conservation of rattlesnakes or tolerance of snakes in the wild when an individual is personally impacted. More research needs to be done, especially on the role that affect, and fear play in predicting tolerance of rattlesnakes. However, respondents with higher domination scores were more likely to perceive greater risks associated with rattlesnakes and therefore desired lower populations. Wildlife value orientations are rapidly shifting from predominantly use-based domination values to more tolerant mutualism-based values (Manfredo et al. 2016, Manfredo et al. 2020), and our model indicates that this could result in Ohio residents being more open to the idea of timber rattlesnake populations being preserved or even increasing. This framework can serve as a critical a starting point for understanding, and ultimately combatting, resistance to the conservation of venomous snakes.

Literature Cited

Agras, S., D. Sylvester, and D. Oliveau. 1969. The epidemiology of common fears and phobias. Comprehensive Psychiatry 10.2, 151–156.

Bruskotter, J.T. and D.C. Fulton. 2012. Will hunters steward wolves? A comment on Treves and Martin. Society & Natural Resources 251:97–102.

Bruskotter, J. T. and R. S. Wilson. 2014. Determining where the wild things will be: Using psychological theory to find tolerance for large carnivores. Conservation Letters 7.3:158–165.

Bruskotter, J.T., A. Singh, D.C. Fulton, and K. Slagle. 2015. Assessing tolerance for wildlife: clarifying relations between concepts and measures. Human Dimensions of Wildlife, 20.3: 255–270.

Ceríaco, L.M.P. 2012. Human attitudes towards herpetofauna: The influence of folklore and negative values on the conservation of amphibians and reptiles in Portugal. Journal of Ethnobiology and Ethnomedicine 8.8, 1–12.

Chan, Y. S., R. C. Cheung, L. Xia, J. H. Wong, T. B. Nq, and W. Y. Chan. 2016. Snake venom toxins: toxicity and medicinal applications. Applied Microbiology and Biotechnology 100.14:6165–6181.

Christoffel, R.A. 2007. Using human dimensions insights to improve conservation efforts for the

eastern massasauga rattlesnake in Michigan and the timber rattlesnake in Minnesota.

book, Michigan State University.

Decker, D.J. and K.G. Purdy. 1988. Toward a concept of wildlife acceptance capacity in wildlife

management. *Wildlife Society Bulletin (1973-2006)*, *16*(1), pp.53–57.

DeLoache, J.S. and V. LoBue. 2009. The narrow fellow in the grass: human infants associate

snakes and fear. Developmental Science 12:1, 201–207.

Forrester, J. A., T. G. Weiser, and J. D. Forrester. 2018. An update on fatalities due to venomous

and nonvenomous animals in the United States (2008–2015). Wilderness &

Environmental Medicine 29:36–44.

Gibbons, J. W., D. E. Scott, T. J. Ryan, K. A. Bulhmann, T. D. Tuberville, B. S. Metts, J. L.

Greene, T. Mills, Y. Leiden, S. Poppy, and C. T. Winne. 2000. The global decline of

reptiles, déjà vu amphibians. Bioscience 50.8:653–666.

Hu, L. and P.M. Bentler. 1999. Cutoff criteria for indices in covariance structural analysis:

conventional criteria versus new alternatives. Structural Equation Modeling 6:1–55.

Inskip, C., N. Carter, S. Riley, T. Roberts, and D. MacMillan. 2016. Toward human-carnivore

coexistence: understanding tolerance for tigers in Bangladesh. PloS one 11.1:e0145913.

Jacobs, M.H., J.J. Vaske, and J.M. Roemer. 2012. Toward a mental systems approach to human

relationships with wildlife: The role of emotional dispositions. Human Dimensions of

Wildlife 17.1:4–15.

Kansky, R., M. Kidd, and A.T. Knight. 2016. A wildlife tolerance model and case study for

understanding human wildlife conflicts. Biological Conservation 201:137–145.

Keener-Eck, L.S. 2017. Human Dimensions of Timber Rattlesnake (*Crotalus horridus*)

Management in Connecticut. Master's book, University of Connecticut.

Keener-Eck, L.S., A.T. Morzillo, and R.A. Christoffel. 2020. Resident attitudes toward timber

rattlesnakes (*Crotalus horridus*). Society & Natural Resources 33.9:1073–1091.

Kellert, S.R. 1985. American attitudes toward and knowledge of animals: An update.

In Advances in animal welfare science (eds. M. Fox and L.D. Mickley) 177–213.

Springer, Dordrecht.

Kellert, S. R., and M. Black, C. R. Rush, and A. J. Bath. 1996. Human Culture and Large

Carnivore Conservation in North America. Conservation Biology 10.4:977–990.

Langley, R.L. 2005. Animal-related fatalities in the United States—an update. Wilderness &

Environmental Medicine 16.2:67–74.

Levin, T. 2016. America's Snake: The Rise and Fall of the Timber Rattlesnake. The University

of Chicago Press.

Lischka, S.A., T.L. Teel, H.E. Johnson, and K.R. Crooks. 2019. Understanding and managing

human tolerance for a large carnivore in a residential system. *Biological

Conservation*, *238*, p.108189.

Manfredo, M.J., T.L. Teel, and K.L. Henry. 2009. Linking society and environment: A

multilevel model of shifting wildlife value orientations in the western United

States. Social Science Quarterly 90.2:407–427.

Manfredo, M.J., T.L. Teel, and A.M. Dietsch. 2016. Implications of human value shift and

persistence for biodiversity conservation. Conservation Biology 30.2:287–296.

Manfredo, M.J., E.G. Urquiza-Haas, A.W.D. Carlos, J.T. Bruskotter, and A.M. Dietsch. 2020. How anthropomorphism is changing the social context of modern wildlife conservation. Biological Conservation 241:108297.

Norris, W. 2017. Massachusetts is scrapping its rattlesnake island plan. Boston Magazine. *<https://www.bostonmagazine.com/news/2017/04/20/rattlesnake-island-plan-canceled/>*

Ohio Division of Wildlife (ODW). 2011. Timber Rattlesnake (*Crotalus horridus*). Ohio Department of Natural Resources. *<wildlife.ohiodnr.gov/species-and-habitats/species-guide-index/reptiles/timber-rattlesnake>*

Öhman, A. and S. Mineka. 2003. The malicious serpent: Snakes as a prototypical stimulus for an evolved module of fear. Current Directions in Psycological Science 12.1, 5–9.

O'Neil, M.E., K.A. Mack, J. Gilchrist, and E.J. Wozniak. 2007. Snakebite injuries treated in United States emergency departments, 2001–2004. Wilderness & environmental medicine 18.4:281–287.

Pandey, D.P., G.S. Pandey, K. Devkota, and M. Goode. 2016. Public perceptions of snakes and snakebite management: implications for conservation and human health in southern Nepal. Journal of Ethnobiology and Ethnomedicine 12.22, 1–24.

Prokop, P., M. Özel, and M. Uşak. 2009. Cross-cultural comparison of student attitudes toward snakes. Society and Animals 17, 224–239.

Reading, C. J., L. M. Luiselli, G. C. Akani, X. Bonnet, G. Amori, J. M. Ballouard, E. Filippi, G. Naulleau, D. Pearson, and L. Rugiero. 2010. Are snake populations in widespread decline? Biology Letters 6.6:777–780.

Riley, S.J. and D.J. Decker. 2000. Risk perception as a factor in wildlife stakeholder acceptance capacity for cougars in Montana. Human Dimensions of Wildlife 5.3:50–62.

Riley, S.J., D.J. Decker, L.H. Carpenter, J.F. Organ, W.F. Siemer, G.F. Mattfeld, and G. Parsons. 2002. The essence of wildlife management. Wildlife Society Bulletin 585–593.

Seigel, R. A. and S. J. Mullin. 2009. "Snake conservation, present and future" *in* Snakes, Ecology and Conservation. Comstock Publishing Associates, Ithaca, NY. 281–290.

Slagle, K. and J.T. Bruskotter. 2019. Tolerance for wildlife: A psychological perspective. Human wildlife interactions: Turning conflict into coexistence 85–106.

Slagle, K., R. Zajac, J. Bruskotter, R. Wilson, and S. Prange. 2013. Building tolerance for bears: A communications experiment. Human Dimensions 77.4:863–869.

Slovic, P., B. Fischhoff, and S. Lichtenstein. 1982. "Facts versus fears: Understanding perceived risk" *in* Judgement Under Uncertainty. Cambridge University Press, Cambridge. 463–490.

Slovic, P., M.L. Finucane, E. Peters, and D.G. MacGregor. 2004. Risk as analysis and risk as feelings: some thoughts about affect, reason, risk, and rationality. Risk Analysis 24.2:311–322.

Slovic, P., M.L. Finucane, E. Peters, and D.G. MacGregor. 2007. The affect heuristic. European Journal of Operational Research 177:1333–1352.

Thomson, C.J. 1991. Effects of the avidity bias on survey estimates of fishing effort and economic value. *In* American Fisheries Society Symposium 12.356366.

Treves, A., and J.T. Bruskotter. 2014. Tolerance for predatory wildlife. Science 344.6183:476–477.

Treves, A. and K.A. Martin, K.A. 2011. Hunters as stewards of wolves in Wisconsin and the Northern Rocky Mountains, USA. Society & Natural Resources 24.9:984–994.

Wachinger, G., O. Renn, C. Begg, and C. Kuhlicke. 2013. The risk perception paradox—

implications for governance and communication of natural hazards. Risk Analysis 33:

1049–1065.

Wynn, D. 2012. Timber rattlesnake populations in Ohio forests. Presentation—Timber

Rattlesnake Workshop, Vinton-Furnace State Forest.

Zajac, R. M., J. T. Bruskotter, R. S. Wilson, and S. Prange. 2012. Learning to live with black

bears: A psychological model of acceptance. The journal of wildlife management 76.7:

1331–1340.

Zinn, H.C., M.J. Manfredo, and J.J. Vaske. 2000. Social psychological bases for stakeholder

acceptance capacity. Human Dimensions of Wildlife 5.3:.20–33.

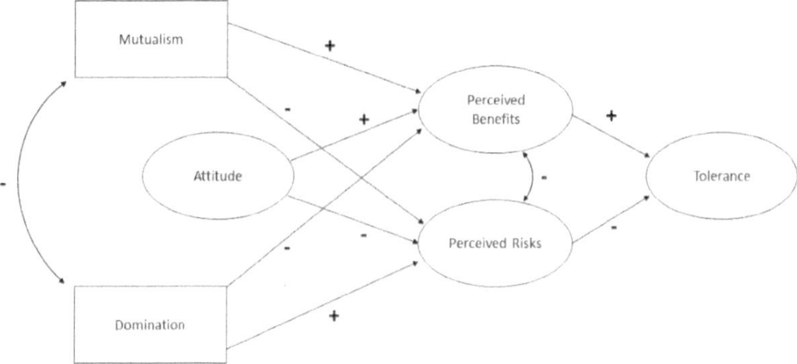

Figure 4.1. Conceptual model for tolerance of rattlesnakes based on our understanding of the latent variables affecting tolerance in large, predatory mammals.

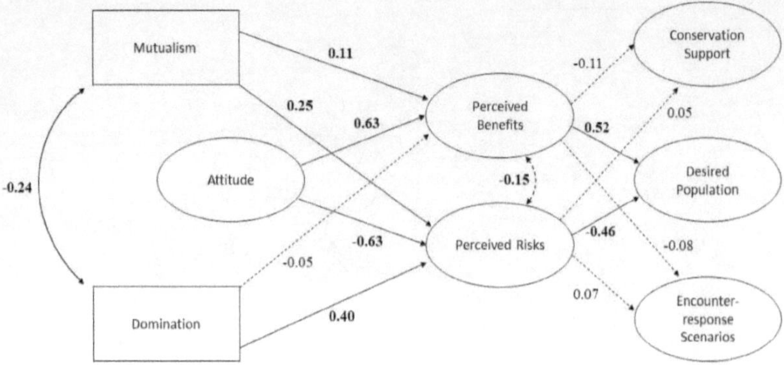

Figure 4.2. Our hypothesized model of tolerance for rattlesnake in Ohio fit to questionnaires from a random sample of residents in the state. The coefficients on each pathway in the model can be interpreted as standardized regression weights. Straight lines represent hypothesized causal pathways and double-headed, curved lines represent hypothesized covariance. We fit three separate models for each of our three measures of tolerance but represent the relationships between perceived risks and benefits and both conservation support and encounter response-scenarios alongside the full model for desired population here.

Table 4.1. Item factor loadings for each of the eight latent variables assessed to model tolerance of timber rattlesnakes (*Crotalus horridus*) by a random sample of Ohio residents. The factor loadings can be interpreted as β from multiple regression output. The Comparative Fit Index (CFI) results present the model fit for the confirmatory factor analysis of each latent variable. The factor analysis results provide the test of unidimensionality in the scales (when CFI scores are closer to 1 the model fits better). * indicates items drawn from the "attitudes toward rattlesnakes" section and operationalized across multiple latent variables.

Latent Variable and Measurement Item Test	CFI	Factor Loadings	SE
Wildlife Value Orientations - Domination			
The needs of humans should take priority over fish and wildlife protection.	0.96	0.574	0.046
It is acceptable for people to kill wildlife if they think it poses a threat to their property.	0.96	0.610	0.045
Fish and wildlife are on earth primarily for people to use.	0.96	0.559	0.046
Hunting does not respect the lives of animals	Removed from analysis		
People who want to hunt should be provided the opportunity to do so	0.96	0.619	0.045
Wildlife Value Orientations - Mutualism			
I view all living things as part of one big family.	1.00	0.791	0.026
Animals should have rights similar to the rights of humans.	1.00	0.712	0.030
Wildlife are like my family and I want to protect them.	1.00	0.783	0.026
I take great comfort in the relationships I have with animals	1.00	0.624	0.035
I value the sense of companionship I receive from animals	Removed from analysis		
Attitude			
To what extent do you like or dislike…rattlesnakes	0.99	0.684	0.029
I am personally interested in rattlesnakes	0.99	0.820	0.021
I would enjoy seeing a rattlesnake in the wild	0.99	0.797	0.022
Even if I never seen one, I enjoy just knowing that rattlesnakes exist	0.99	0.702	0.028
I take pride in knowing that a rattlesnake lives near my home	0.99	0.727	0.027
Perception of Benefits			
Rattlesnakes help to control pest populations	NA	0.639	NA
Rattlesnakes are important to the Ohio ecosystem	NA	0.866	NA

Perception of Risks

Rattlesnakes pose an unacceptable threat to pets	0.95	0.791	0.031
Rattlesnakes pose an unacceptable threat to children	0.95	0.757	0.033
If I knew that a rattlesnake lived near my home, it would decrease my enjoyment of living there	0.95	0.622	0.036
I am less likely to visit a park where rattlesnakes occur	0.95	0.556	0.039
Desired Rattlesnake Population			
Do you think rattlesnake populations in Ohio should...	NA	0.877	NA
What would you consider an ideal population of rattlesnakes in Ohio?	NA	0.719	NA
Behavioral Intentions Toward Rattlesnakes			
You find a venomous snake on your property	1.00	0.550	0.035
You find a venomous snake crossing a road	1.00	0.813	0.035
You find a venomous snake along a trail on public property	1.00	0.819	0.035
Support for Rattlesnake Conservation			
Laws that prohibit killing rattlesnakes	1.00	0.712	0.031
Laws protecting rattlesnakes that restrict a landowner's right to develop private property	1.00	0.667	0.033
Reintroduction of rattlesnakes at sites where they've dissappeared	1.00	0.752	0.029
Introduction of more rattlesnakes into a declining population	1.00	0.740	0.030

www.ingramcontent.com/pod-product-compliance
Lightning Source LLC
LaVergne TN
LVHW041728190726
843493LV00007B/2259